Sonja Englert

Tips and Tricks for Aircraft Homebuilders

First Edition, Bend, Oregon, USA

http://www.caro-engineering.com

ISBN 0-9752984-8-8

Introduction

Aircraft homebuilders are innovative people and invent many practices and techniques to help them reach their goal: to build the best airplane they possibly can. The homebuilding community is unique in that the members freely share tips, lessons learned and advice with others. There are thousands of ideas and neat things that can help you build a better, safer and less expensive airplane. I have collected many of these useful tips and assembled them in this book. They range from tips on how to get started, to construction methods, step-by-step instructions to build certain things, schematics, tips on assembly and flight testing. There should be something for everyone. While many of the tips are my own, learned in over 25 years of flying, working on airplanes and building them, I have also used ideas, descriptions and pictures from other builders. The names of those who have contributed to this book are listed here in alphabetical order. It was not always possible to get the name of everyone, so please accept my apologies if I missed someone. Because most tips are a combination of the input from several builders, I have not listed the names with the description of the tip.

Builders who have contributed tips or ideas:

Barry Edwards
Bill Schertz
Bob Hartunian
Bob Reed
Brian Anderson
Brian Meyette
Bud Newhall
Charlie Kuss
David Morris
Debo Cox
Eric Newton
Hugh Kenny
Jerry Carter
Jerzy Krasinski
John Moran
Kirby and Bob Jefferson
Kirk Harrell
Laurel McKone
Len Fox
Luigi Aldini
Tim Coons
Mark Langford
Michel Therrien
Mike Cowan
Mike Henderson
Nate Wolfe
Orv Krohn
Paul Besing
Ray Clegg
Rick Clarke
Rob Miller
Roger Hirschbein
Rolf Hankers
Ron Koval
Ron Willett
Sam Buchanan
Stefan Vorkoetter
Tom Nalevanko
Wallace Judd
Colin Lamb
Gene DeMarco
Jeff Michaels

A big THANK YOU goes to all of those who have contributed!

Table of Contents

Getting Started

Planning the Work

Read all the build manuals until you understand them before starting to build. To work efficiently, have a building plan. Do not just follow the instructions step by step from the manuals or drawings, but figure out where you can do several jobs from different manuals at the same time - sometimes the tasks are related. Stick to the plan and always have a primary, secondary and tertiary job list for every visit to the workshop. Then, if you have time or glue etc left over, it can be used productively. If you have a very small final job, it may be quicker to do it at the end of other work rather than starting fresh next time. For example if you have a little bit of filler available at the end of a job you can use it for some finishing work elsewhere. Don't stand in the workshop and look at it! Work on it at every opportunity if you want to finish it.

Make a list of everything you can think of that you need to do to build this plane. Date each item on the list as it is completed, but leave it on the list. When a roadblock interferes with the current task at hand, a quick reference to the "To Do" list will get the project moving again, because the list can be used to pull yourself away from the task at hand.

Incomplete items: At any given point in the project, but especially when systems are being installed in the airframe, there are usually a number of unfinished tasks going on at any time. Keep a "To Do" list and do not mark anything off it until that task is torqed, safety wired, double checked... etc. Keep the full list and date each item when it is complete. If you have to remove a bolt or take off a safety wire, take the date off of the completed task.

Keep a daily log of time on task. Just a short note of what was being done. Then when frustration sets in or it feels like you are not getting anywhere, a look at the log and "To Do" list will give you the reassurance that there is progress being made.

Try to do *something* on the project every day - even if it is only for a few minutes.

If people come to visit, after a few minute tour, get back to work while talking to them. If they insist on staying, put them to work.

Take a lot of pictures, of your own project as well as of other builder's projects and airplanes in general. You do not have to sort them but you can learn a lot on a rainy day. Digital photos cost nothing these days.

Building Time

Once you start building, you will find out what works best for you. Some people are most effective working about a 3-4 hour building session a day, but maybe not every day. Others may be used to doing 10 hours straight or just one hour in the evening on a regular basis. Each builder has a particular set of constraints to work with and effective working hours differ from builder to builder. It is really important to know your working habit and limits.

When kit plane factory quotes you what it calls 'build time', they usually mean time you spend working on the plane. It does not include what builders call 'head scratching' time. This is time we all spend looking at the manual, watching the video, calling factory support, and going through the Aircraft Spruce and Specialty catalog. Take time to think about what needs to be done next. Read ahead and see how each process will come together in the end. Then mentally plan out the work for the next session and have some rough idea of what you are going to do in what order. If necessary, order supplies and tools you might need ahead of time. By getting the planning out of the way, you can concentrate on getting the work done and use the limited shop time effectively.

- Arrange your work so that you do a task which requires curing or drying time at the end of your building session. Then you are less likely to touch it by accident or get dirt on it.
- Finish one task and give yourself some time to study the next one before jumping in.
- Keep up a regular building schedule, but do not get so carried away that you neglect everything else in your life. Balance building with other aspects.
- Don't build when you are tired, you are much more likely to make mistakes, which you will discover the next day and waste time redoing it.

Get Advice Early

It is always better to learn from other's mistakes rather than having to make them yourself. So read this paragraph and decide for yourself. One builder described a lesson he had to learn the hard way.

The seatback is the very first part he attempted to build of his airplane. It is a flat, moderately sized composite layup that is a good first project to let you learn on. The builder reports he had to re-do the part due to the fact that he really did not get up to speed on the peel-ply process and working out the bubbles before he started. As a result, his first part ended up with too many defects and he had to cut it up. While this might at first seem sad, it was an excellent lesson on many levels. It got him involved with other builders in the area because he immediately knew that the first piece was NOT how he wanted his project to go. A visit from another builder, a lesson in the best ways to learn the skill, consistent products and some new material from the supplier later, and he had a seatback that was excellent!

ADVICE - Get help early to aid in your confidence and consistency. The remainder of the parts this builder made turned out great because of this first lesson.

Understand the Concept of Your Kit

If you are building a mature kit that is known for its exact nature and user-friendliness, perhaps following the directions and drawings will be all it takes. However, if you decide to build an early kit or an immature kit, you should understand that you might have to "build to fit". That means that there are some parts you receive that are likely to change and that you need to do your constructions around them. There may be errors in the instructions or drawings. Do not blindly start making parts because these may never fit together. Build defensively, cross-check and never assume that components will just fit together.

Shop and Tools

Heating the Shop

If you plan on working in winter as well as summer, you should think about how you want to heat the shop. Depending on the size and insulation, you may need anything from 6000 W to 2 kW. If you do not want to heat a large space, you can partition the space you need to work in and just keep it warm. If you spend time irregularly in the shop, it will hardly be worth it to keep it warm all the time. Then you should select a heater which can heat it up in a short time. Here are the methods used by some builders:

- Used oil fired hot water furnace. It was piped to a car radiator with a fan blowing through it. It is fed by a 55 gallon oil drum behind the garage that the oil company fills when they fill my house tank. Hot air comes out of the radiator at about 140°F.

- An unvented oil heater works well and burns very clean. CO levels are well within the acceptable range when running the heater. Because it only burns cleanly when it is at its operating temperature, there may be some smell when it is lighted and warming up. You absolutely have to take it outside before turning it off. I neglected to do that once and it took weeks to get rid of the smell.
- Hot air furnace (the ones from mobile homes are particularly compact).
- Wall mounted vent-less propane heater or natural gas convection heater, which can be purchased from McMaster Carr (60,000 BTU/hour output). A convection heater is more suitable to heat the entire volume of air of your shop evenly as opposed to a radiant heater which primarily heats objects in its path.
- Circulating hot water heat in the floor of the shop. Heat source is 30 gallons domestic water heater about 30,000 BTU, propane fuel. Circulation temperature typically about 85 F. 500 ft of 3/4 inch polybutylene tube in floor in two counter-circulating loops. Each loop goes out to two of the perimeter walls first. A small circulation pump runs continuously, drawing under 1 Amp.

Air Compressor

For most projects it is somewhere between nice to have to absolutely essential to have an air compressor. Many tools are powered by compressed air, it can be used to blow dust of a surface and it is needed for spray painting. The size of your compressed air storage tank is as important as the compressor itself. It really depends on what you are using it for, but in general, larger is better. The more volume you have available, the less often will the compressor have to come on. Because the compressor is noisy, it is desirable for it to either run infrequently or have it located far away from where you are working. For work requiring a lot of air like sandblasting, a compressor and tank suitable for operating tools is inadequate. An air compressor is an important part of your shop. Get a good one, not the smallest, cheapest one which may cause you headaches and cost you more in the long run. The same applies to hoses and fittings. Rubber hoses age and eventually leak or blow out (hopefully after your project is finished). Learn about the options first and evaluate them before you put your money on the table.

Vacuum Pump for Free

If you already own an air compressor, you can convert it to a vacuum pump for vacuum bagging composite layups. Remove the air intake filter. Use threaded pipe about 3 inches long, cover the threads with Teflon tape and screw it into the cylinder head air intake. Then take a wooden dowel about 1.5 inches in diameter, 3 inches long and drill a hole into it length wise about the size of the little attachments you clip compressor tools to (about 1/4 inch) to act as an adapter. Then screw this adapter in. Clean out any dust or chips to prevent them from entering the compressor. Buy a vacuum gauge (for auto brakes for example) and connect it inline. Slide the dowel into the threaded pipe and wrap the whole thing with duct tape to prevent leaks. After attaching a hose, turn the compressor on and watch it pulling vacuum. The theoretical value of the vacuum is 30 inches of mercury (inHg), but you can be satisfied if you get over 28 inHg. The compressor will still function like a compressor. When you need to pull vacuum, just move the hose to the other attachment. You may want to remove the oil fill plug so it does not suck oil into the cylinder.

A different approach consists of removing and modifying the reed valve setup in the pump head, and then installing it so that it sucks instead of blows. As such, you do not have to redo any plumbing and the tank now acts as a vacuum plenum, which you can close off from the hose with a simple valve. Then, when the laminate is done and bagged, all you do is open the valve so that you get an instant vacuum with a high rate flow. The pump then keeps the vacuum magnitude in the tank.

Machine Set Up

Many machines have features for set up for a specific task. Examples would be the angle of a drill press table or sander, stops or guides on a band or table saw, stops or bend radius on a brake. This can help you accomplish a certain job with more precision or for repeated operations. Once you have set the machine up, check how it performs with a piece of scrap material. Measure it to verify this is really what you wanted. Often you will find that a further adjustment is needed because of flexing of support, machine, blade or spring-back.

Magnets as Clamps

Have you ever had to reach around corners, way down into hard to access spaces where you needed to clamp a part? Consider magnets as clamping devices. You need two of them, one on the outside and one on the inside to clamp two pieces together. The thicker the parts are, the lower is the clamping force because magnetic force rapidly diminishes with distance. For some really serious clamping force, use super strong rare earth magnets. Be careful! Neodymium alloy magnets are strong enough to be dangerous to work around. Big ones can easily break or crush fingers if they take off and snap together.

Clamping Small Parts in a Vise

Sometimes it is necessary to clamp a small part in a vise on one end. Unfortunately the edges of the vise do not stay parallel and the part is not held securely. When pressure is applied to the part, in which you wanted to drill a hole, it will rotate down. To avoid this problem, simply insert a second part of the same thickness at the other end, this will keep the clamping edges parallel.

Scale in a Bag

An electronic scale is the most useful tool for measuring resin in the correct proportions. Unfortunately it seems to be impossible to pour resin into a cup on the scale without spilling some. I put the scale in a plastic bag to protect it. I can push the buttons and read the display through it. When it becomes too sticky, I replace the bag.

Procedure to Sharpen Scissors

Scissors used for cutting fabric (glass, carbon etc.) work fine for a while. But because the fabric material is very hard, the blades dull quickly. The results are frayed fibers, uneven cuts and blisters on your fingers from the hard work, especially if Kevlar is involved. Sharp scissors cut easily and cleanly. Unless you are able to sharpen your scissors frequently, cutting fabric can quickly become a frustrating job. To sharpen scissors, you will need a grinding wheel, which can also be used to sharpen drill bits and other things. Open the scissors wide and hold one blade against the rotating wheel at the angle shown in the sketch. Start as close to the hinge of the scissors as possible. Slowly draw the blade across the wheel towards its tip while maintaining the same angle. You should just see the sparks coming over the top of the cutting edge. If they stay below the edge, you are not reaching the edge. Look at the edge; it should be at the same angle the full length of the blade. The old surface should have been ground away. If any of the old surface at the cutting edge remains, repeat the process. Once you have sharpened both blades and close the scissors for the first time, close them by pushing the blades apart with your hand so that they do not touch until the blades are fully closed. Then open them while pressing the blades together. This removes the slight ridge on the cutting edge which is the result of the grinding. This can also be done with fine sandpaper or a stone drawn over the inside of the blade. If you do not remove this ridge, the scissors will not cut well and dull very quickly. Now the scissors are ready to use.

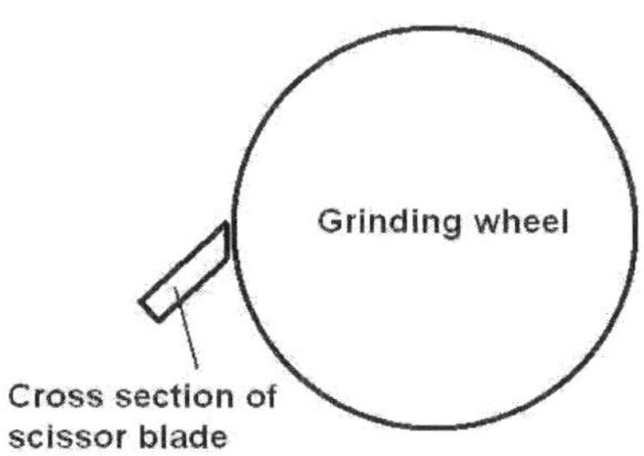

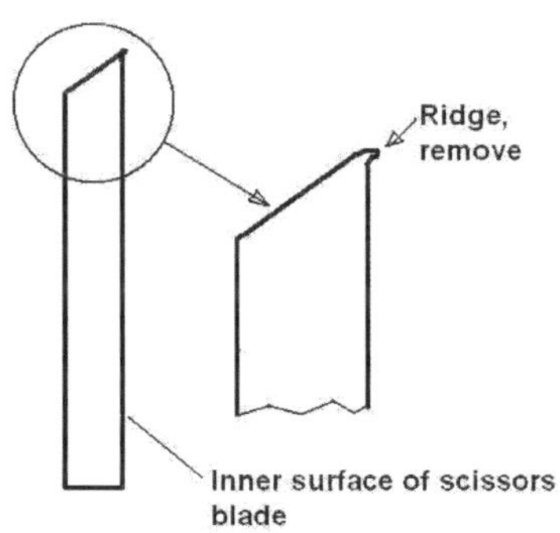

Spiral Tubing Cutters

Spiral tubing can be used to keep wire bundles together or as anti-chafe covering for tubing. While you can buy spiral tubing in various sizes, you could also make it yourself if you have surplus regular tubing available. To cut the tubing, for example polyethylene, into a spiral, you need to make a very simple little tool. Take a scrap of wood and drill a hole near the edge that is slightly larger than the tubing you wish to cut. Saw a thin kerf in the edge of the wood, to intersect the hole at an angle, bond a razor blade into the slot such that it protrudes into the slot just enough to cut through the wall of the tubing and the tool is completed. To make spiral tubing, start feeding tubing into the hole with a twisting motion while pulling it out the other side. If you have different diameter tubing, make a tool for each size. The angle of the razor slot establishes the spacing of the spiral cut. You will have to experiment a bit to get what you want.

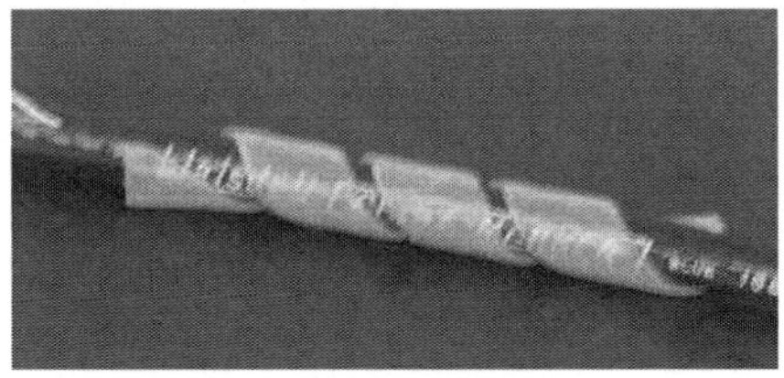

Drill Press

Maybe you can build your airplane with only a hand drill. But I found that a bench model drill press is very useful. A drill press is especially valuable for cutting large, accurate holes. For small holes, a high speed hand drill works fine. When you are shopping for a cheap one, watch for a few things. The table should not flex down when you put a lot of pressure drilling something hard. The speed should be adjustable. This is typically done by moving the drive belt from one pair of pulleys to another. The table on cheap models slides up or down for adjustment. This is fine if you do not mind the extra work. It is more convenient if the table can be cranked up and down with a hand crank.

Depth Stop for Drill Press

The depth stop on an inexpensive drill press can be so awkward and time consuming to adjust that it is seldom used. If it requires winding two nuts up and down, you are probably thinking it is not worth the effort. The picture shows an alternative approach which you can build yourself, which allows fast and easy adjustments of the stop. The concept is the same as the cotter used to lock a lathe tailstock. It takes only a fraction of a turn to lock or unlock and is very secure when locked. You may find that you use the depth stop far more than anticipated when you have this modification.

Construction is simple and you can use whatever material you have available. You will need a rod to replace the original threaded rod. A rectangular metal block is used as the stop block body. A cotter is made from 3/8 inch round (about 3/16 inch longer than the stop block body) by end drilling for a 10-32 tap, clearance drilling just over half the length, then tapping.

The stop block is pilot drilled to provide the intersecting holes when drilled to size, where the hole for the stop rod is positioned so the block will be about 1/16 inch from the scale of the drill press. The pilot for the 3/8 inch hole is placed so the center clearance hole will have about a 0.050 inch wall in the bottom of the cotter. The cross-holes are drilled to size, carefully, (shaft hole first) because of the interrupted cut part way through. The shaft is put in place and drill shanks are used to try in the 3/8 inch hole to measure the clearance; this shank size is subtracted from 0.375 inch to find the depth of the cotter. The 3/8 inch round is side milled with a 15/64 inch cutter at the center to the appropriate depth for the cotter, then cut in half at the bottom of the cotter with a 0.015 inch slitting saw. As the knob you can use any knob you may have available. It is drilled for a press fit on the head of a 10-32 socket head bolt. Cross drill the stop rod but if you have a suitable nut, a longer thread would be another way to lock it in place. With the rod in place and the cotter positioned to match the hole for the rod, run the stop block onto the rod and tighten the knurled knob to lock the block to the rod.

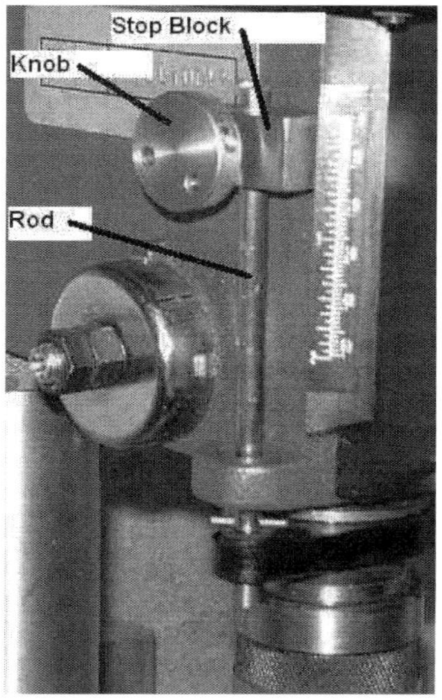

When not in use, the stop block is positioned so the top edge can be used with the scale for depth readout. To use as a depth stop, loosen the knob and lower the drill to the desired depth, push the stop block down against the contact and tighten the knob to lock the cotter. It takes little torque to lock it solidly. Alternatively, to drill to a depth, lower the drill point against the item to be drilled and put a spacer equal to the desired depth between the stop block and its contact point, push the stop down against this spacer and tighten; remove the spacer and the depth is set.

Considerations for a Fuselage Turning Jig

A fuselage turning jig enables you to rotate the fuselage by yourself so you can work on it at different angles. Depending on the size of the fuselage, it may or may not be practical to build such a jig. Consider the height of your shop first. With the turning axis level and the fuselage upside down, with the vertical tail pointed down, how high would the fuselage belly be? Can you still reach it to work on it? Will it still fit in your shop? Would it be sufficient to turn the fuselage sideways?

Once you have determined how far the fuselage will be turned, examine the fuselage for hard points to attach it to a jig. In front, the engine mount hard points are probably the most suitable ones. On the tail this might be more difficult. Does the tail end have hard points for the attachment of the horizontal tail? If not, the best solution is usually to fit the jig around the thinnest part of the fuselage itself, at the location of a bulkhead so that the skin is not overstressed.

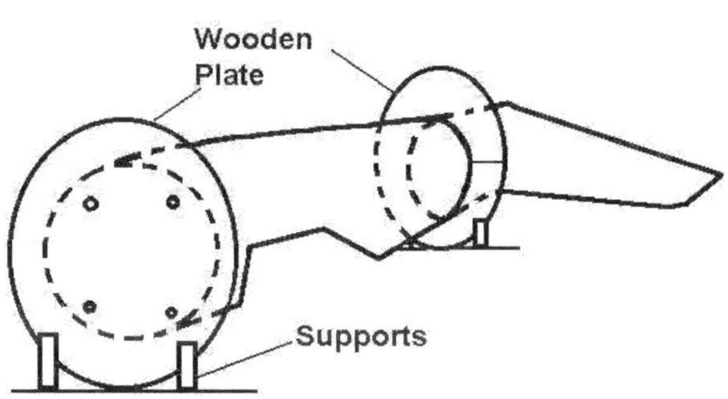

A very simple jig consists of round pieces of plywood. The fuselage can be rolled around on the floor to any angle only limited by the vertical tail. It is low enough to easily work on. The height is determined by the diameter of the wooden plates. Support legs which are bolted on, hold the fuselage at the desired angle.

If your fuselage is small and needs to be raised up to work on it, consider the following jig. The support frame is spaced to fit on a saw horse. The pipes can be metal, or plastic if the fuselage is light enough. The pictures show a turning jig constructed using 2-1/2 inch PVC plumbing pipe and fittings. These are assembled and glued together with a framework of wooden boards. The two vertical legs are made to slip down on either side of the sawhorse. There is a spacer used to lock the pieces together and keep the jig from sliding too far down on the sawhorse. The holes are used for a nut and bolt which can be tightened to lock the jig at any desired position.

The tail end of the fuselage can simply be supported on a saw horse.

Another possibility is buying an engine stand and fabricating some 1/4 inch steel angle to fit the engine mounting holes on the firewall. This stand allows you to move the fuselage around the shop and you can rotate it when working inside or on the bottom of the fuselage.

Consider if you really have to build and attach the vertical tail before you have done most of the work on the fuselage like gear, etc. Once you build the vertical tail, rotating the fuselage becomes much more difficult.

Wing Turning Jig

When you prepare to paint a wing, keep in mind that for best results it is important that the stream of paint remain continuous as you move across the surface area. The problem that comes up is how to keep that stream of paint continuous as you move from the top to the bottom of the wing. A solution which has worked well for some builders is described here.

Begin by constructing two round wooden discs about 24 inches (or a size appropriate to your wing) in diameter and cut them in half. Using a band saw, notch each piece of the half round to fit the dimensions of the spar. After you place the two half rounds over the spar, use steel plates drilled with 4 holes to bolt the two half rounds together. This way, a wheel is placed on either side of the centerline of the wing spar as shown in the left sketch.

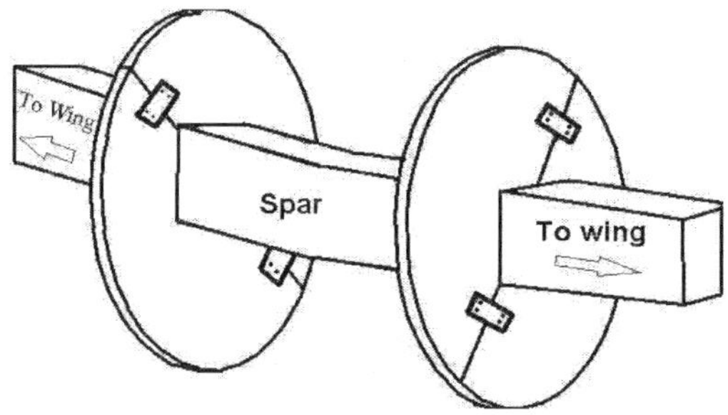

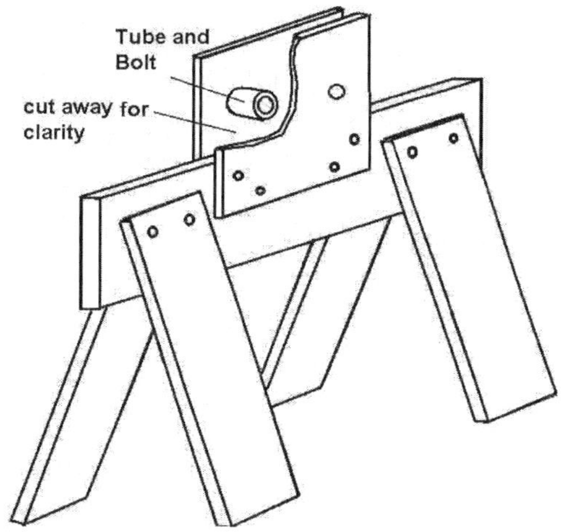

Once the half rounds are completed and attached to the spar, a jig is fabricated to hold each of the half rounds so that you can rotate the wing through the full 360 . This jig can be built from two sawhorses.

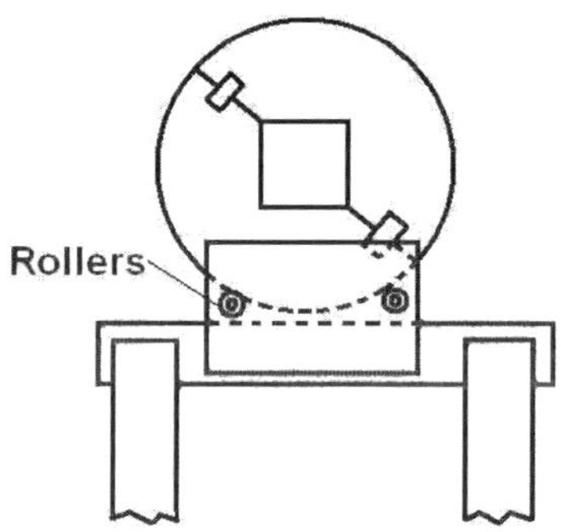

Bolted to each sawhorse are two plywood pieces that are for example 10 inches by 16 inches in size. The half round is nested between each of the two pieces of plywood to determine the location for a pair of rollers. The half round will be supported by the two rollers and not directly by the sawhorse. Drill two holes and install a bolt at each location. Slipped over each bolt is a sleeve made of 1/2 inch tubing (right sketch). The purpose of the sleeve is to improve the ease by which the wheel can be rotated in the nest as you reposition the wing. It is important that the nest rollers maintain enough friction to keep the wing in position at locations that are not natural balance points as you will want to position the wing at several angles as you move through the painting and sanding process.

Cutting Large Holes

There are a number of tools used to cut large holes (larger than regular drills). The choice of tool to use mainly depends on the material and on the size of the hole. These are my choices:

- Step Drill
  Good for smaller holes and softer materials. Dulls quickly when used on composite.
- Hole Saw
  Cuts only one size hole. Available with saw teeth for wood and metal or carbide coated for composite. Comes with or without center drill. Needs to have holes or slots so the disc you cut out can be removed.
- Spade Bits
  Best for drilling metal, limited in size. Lubricate with oil while cutting.
- Fly-Cutter
  Single lathe type bit, variable hole fly-cutter, for use in all materials up to mild steel. Double bit versions are available for more stability and faster cutting, but are more difficult to setup. This it is a very dangerous tool and should only be used in a drill press. Always firmly clamp the part to the table so it cannot move. As a precaution, leave the belt tension of the drill press as loose as you can

and still turn the cutter. That way if it catches and throws the part or hits something (like your hand), it will do less damage. It cuts all materials but is best with metal. Use slow speed.

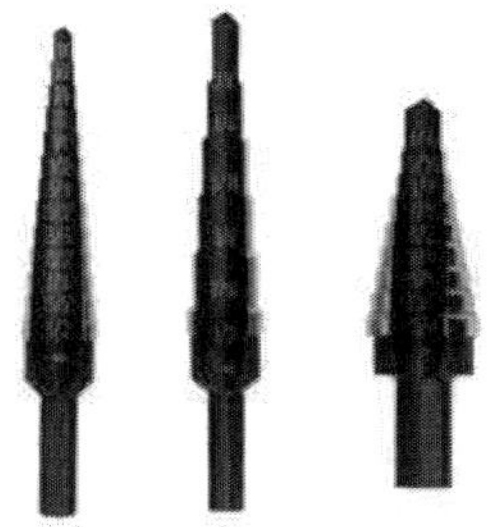

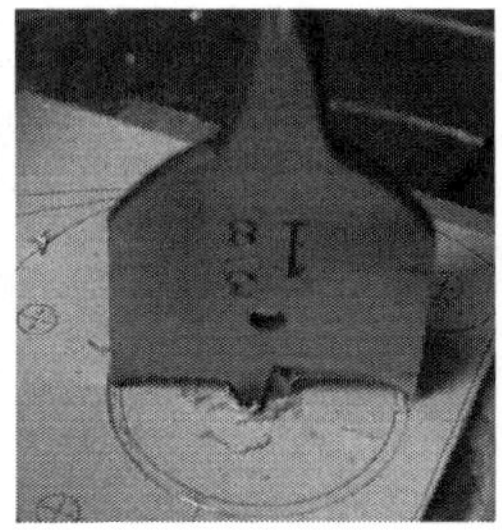
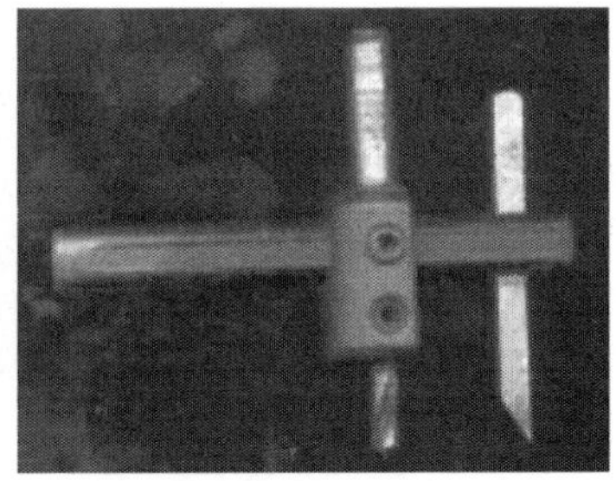

Step Drills Hole Saws Spade Bit Fly-Cutter

Protection of Power Chords

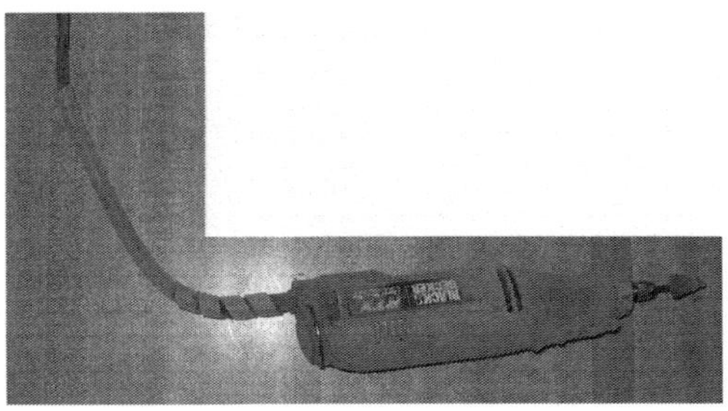

If you notice the power cord of your favorite, most used tool (a Dremel tool in my case) always getting kinked in the same place, the conductor will fail sooner or later. This happened with the tool in this picture. To prevent this, stiffen up the affected section with spiral wrap. The spiral wrap should be smaller in diameter than the chord so it sits tightly on the cord once it is installed.

Flat Surface

If you need a flat, smooth surface, for example to do a layup on, think about using an old glass window pane. Wax it to release it. It should be thick enough not to flex. If it is still in a frame, it will hold securely and straight.

If you are building a table with a flat and straight surface, one of the best surfaces is a 3/4 inch laminate coated pressed board obtainable from many builder's supply stores at 4 x 8 feet. Construction methods for the table vary from builder to builder, but remember the purpose of your bench. To assemble a wing or fuselage, you may wish to keep the height to around the knees, to work on skins and manufacturing smaller pieces you may wish to keep it at a height where you are not constantly bent over.

Dust Removal Table

If your project involves a lot of sanding, you will end up with a lot of dust in places that could really do without it. The easiest way to avoid dust in your shop and on yourself is to move the part outside and have the wind or a fan blow it away from you. Unfortunately this is not always an option.

To lower the dust load in your workshop, you can also build a filtering sanding table that removes the dust from sanding before it is distributed on everything. To use it, put anything that needs sanding on it and turn on the fan. The air and the dust are drawn downward through the perforated table surface into the table where the air is filtered and expelled through the side. To build this, you will need:

- One 4 ft x 8 ft sheet of plywood, 3/4 inch thick
- One 2 ft x 4 ft sheet of perforated hardboard, 1/4 inch thick. The holes should be about 1/2 inch in diameter.

- One squirrel cage blower/motor from an old furnace (free - just talk to an HVAC repairman and ask for one)
- Some miscellaneous scraps of wood.
- Four 20 inch x 20 inch pleated furnace filters, 1 inch thick.

An overall view of the unit is shown in the photo. The 3/4 inch plywood is used to construct an open topped box that houses a furnace blower. You can see the blower exit, a switch plate for turning the table on, and a couple of plugs for plugging in a sander. The top is the perforated board. You can paint the top with a polyurethane varnish or resin that hardens and toughens the perforated board and makes it smooth. The black sheet that is sitting on the back half of the top in the left picture is a porous rubber mat that lets the sawdust go through, and prevents things from skidding around the top.

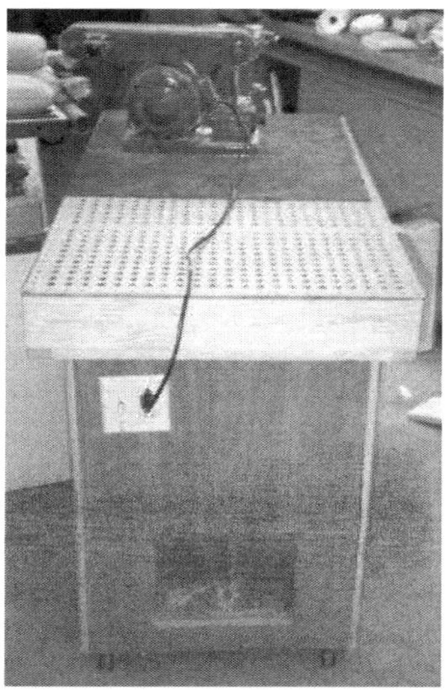

You can see a belt sander on the table. With the table blower running, the dust from the belt sander just disappears into the table and does not enter the shop. The other photo shows the unit with the top removed. You can see the furnace filters in place, and the cross braces for the perforated top. The rim around the perforated top was sized to get a full 2 ft x 4 ft surface, and still build the box enclosure with only one sheet of plywood. The filters are 1 inch thick, and you can stack two on top of each other to get a 2 inch filter. This photo shows the internal workings of the unit. The blower is just mounted so that the exhaust blows out of the box, creating suction below the filters. The filters rest on a wooden ledge. You can see some material on the filter medium. Rather than frequently purchasing new ones, you could periodically vacuum off the accumulated dust from the filters. You can even turn on the table when not working on sanding because it acts like a fan for the shop. If you have done another operation that generated a lot of dust in the shop, this unit will remove it from the air very quickly. You can mount it on castoring wheels so it can easily be moved around to the dustiest places.

Weights

Often there is a need to hold things down, for example when you are gluing skins, spars or other parts. If the part is large, you may need a lot of weight. Instead of buying shot bags, you can use a number of less expensive things. You can fill ziplock bags with sand, or if that is too small, sew bags from sturdy cloth and fill them with sand. That has the advantage that you can make them any shape and size you need them, long and narrow or short and wide. Old clothes can be used to speed up the sand bag manufacture: cut off sleeves and pant legs, sew one end closed, fill them and tie them at the top. Another solution is to collect empty jugs and fill them with water. Or fill up orange juice cartons with a mixture of sand and lead weights you beg from a tire change shop. Use the type of carton with a screw cap near the top or you will get sand leaks. Wrap the carton in duct tape, weigh it and write the weight on the carton for future use.

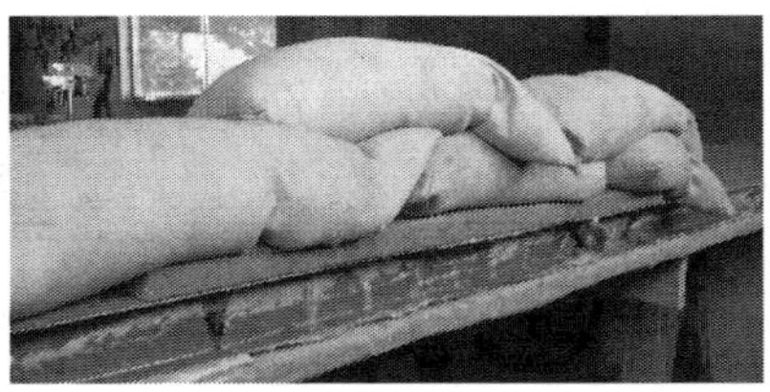

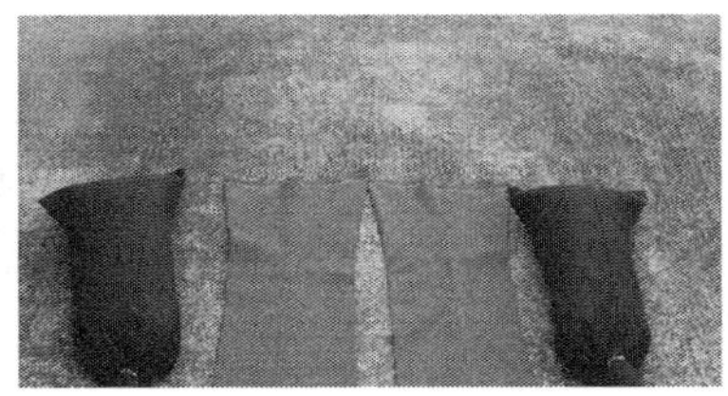

Level the Wings

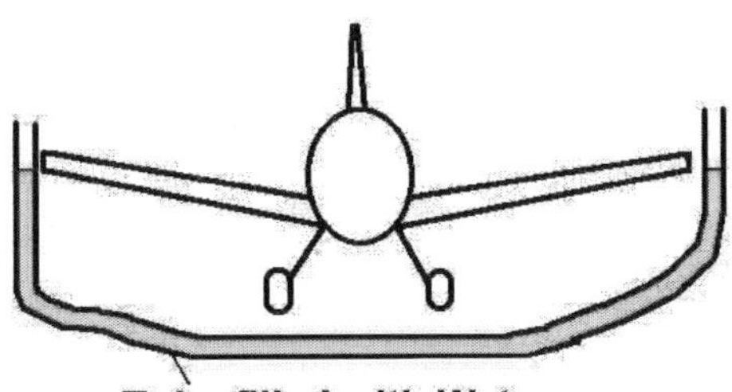

When you need to build something perfectly level, there is a very simple tool to help you. For example you are attaching wings to the fuselage and want the dihedral on each side to be exactly the same. There is no need to measure the distance from the wing tip to the floor and hope that the floor is level. Get a piece of clear plastic tubing a bit longer than your wing span. It needs to be of a fairly large diameter, the longer the larger to not restrict the flow. Fill it with water and hold the ends up. First, hold them close together to check that the water levels properly. Then hold them up at each wing tip. The ends of the tube must be open. The water level will equalize and be exactly at the same height left and right. Adjust the wing tips to the water level and they will be exactly the same height. This concept for leveling distant parts can be used anywhere else as well.

Bicycle Spokes

These have to be the most useful ad hoc item in a shop and they can usually be had free. They can be bent into a fixture, used as hangers or used as a flexible poor man's drill bit. I am sure once you have some, you can come up with many uses for them. If you do not have an old bicycle wheel, go to a bicycle shop and ask for an old wheel or spokes, which you could have for free.

Table as a Jig

Building a flat table may be the first thing you need to do when starting your project. In many cases this table is utilized as a jig for building sections of your airframe. This table must be flat and level (both side to side and lengthwise). Having the table flat and level will help greatly later on while trying to get everything square and plumb. The top of the table should be painted flat white with latex house paint and a roller. This will allow you to see lines and markings easier. When you are done with one part, you can re-paint the table top to get rid of the old lines. Once the paint is dry, you can snap a chalk line down the middle of the table full length and trace it with a Sharpie marker and a long straight-edge to make it more permanent. This line can serve as the centerline of the components you are building.

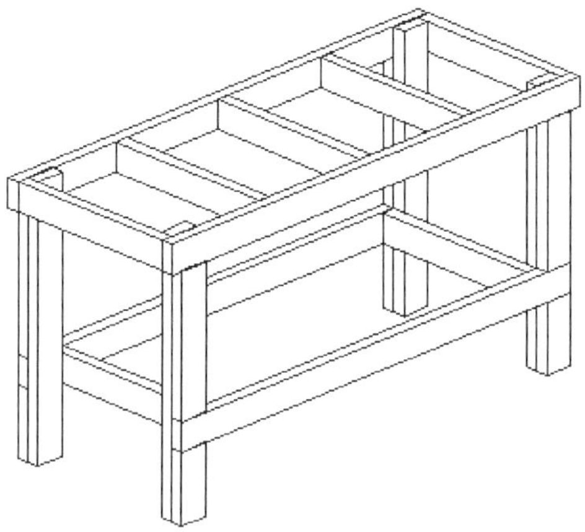

This is an example of a table's structure. The top and the shelf underneath are covered with plywood (not shown). The structure is glued as well as screwed together. Because hardly any floor is perfectly level, you will either need to shim the legs of the table or add leveling screws to the legs. Instead of building one long table, you can more easily build several smaller tables and bolt them together. After the project is completed, you can take them apart and use them for other jobs.

If you need additional table surface but don't have the space for another permanent structure, use two sawhorses and a sheet of plywood for a quick setup.

Wing Jig

For an airplane to fly straight, first of all it needs to be built straight. This is a particular challenge if your airplane kit consists of pre-molded composite skins. Before the skins are bonded into a rigid assembly, they are very flexible. It is very easy to build warped parts from those flexible skins unless they are supported properly by a jig while they are bonded. A bonding jig is important enough to do it right and make sure it is built to do its job.

To build a jig, use materials that start out straight and true and will probably remain that way. The humidity in some areas can cause just about anything from wood to warp. Pre-finished 12 inch by 48 inch shelves of 5/8 inch (16 mm) thick MDF boards are suitable material. You will need to have available the shape of the contour of the part you are building the jig for. Ask the kit supplier for the pattern if you do not have it. Cut a number of ribs to roughly the shape of the pattern. These ribs will support the skin. How many ribs you need depends on the size and stiffness of the part. The spacing of the ribs for the relatively stiff sandwich structure of a wing skin should not exceed two feet. Shape one board to the exact size of the pattern. Once you are satisfied with the initial rib, use it to produce more exact copies (assuming the wing planform is constant chord). The best method for duplicating a wood part is to use a router. The whole process should not take more than an hour. All complete ribs are clamped together and checked for true match. Use the same process to cut the slots for the center support.

For the rails, select the best and straightest boards you can find. The rails provide the support on both sides and in the center. Place the rails together and mark the position for the ribs on all three rails at the same time. Use wood blocks to provide the exact position of each rib.

Assemble the jig on a work bench which has been leveled, with dry wall screws. Everything needs to be repeatedly checked to be level and square. Use wood blocks to position and anchor the whole jig to the top of the workbench. All the ribs are screwed into place with anchors liberally placed to hold the jig in the proper position.

This picture shows the position of the jig on the work bench. It should last through both wings without warping. The best jig cannot do its job if the wing skin is not firmly attached to it. In order to hold the wing skin down on the jig, bond it to the jig at each rib, the trailing edge and leading edge with blobs of Bondo, or if something stronger is needed, with micro or flox. Make sure to not sand or clean the wing skin so that the bond can easily be broken when the wing is complete. You may need to clamp or weight down the skin until the adhesive is cured to hold it place. Use a long straight edge to hold down the trailing edge full length.

Removal of the wing from the jig should not be a problem. The flox should have held tight through the construction process and release with a bit of friendly persuasion. Lift the wing from the front and slowly pop the bonds loose.

Fuselage Mobility in the Shop

You have your fuselage on wheels but in the limited space you have in the garage it always seems to be in the way and hard to move. Does this sound familiar? Get a pair of cheap dollies and put them under the wheels. This allows easy sideways movement while fuselage is on its gear. A small 3-wheel dolly used to move snowmobiles in the garage works great under the tail spring.

Drilling and Supporting Large Pieces

If you ever have to drill spar caps, you will probably be wondering about the best way to drill perfectly aligned holes into them and still support the remaining ends so no bow is built into the spars. You may have seen where other guys set their drill press beside their work table and swing the head around, but that is a little like driving a car from the back seat. It is hard to make sure everything is perfectly aligned, and for this critical drilling operation you needed to be more in control.

Here is a solution. Use a hole saw to cut a hole in the workbench so you can stick the post of the drill press up through it. The head of the drill press should come off with two simple Allen screws, so it should not be difficult to remove it, then stick it back on. The drill press is resting on a solid small table under the bench. You can even get the normal adjustable drill press table cranked up and braced against the bottom of the workbench - further stabilizing it. After leveling and making sure everything is square, cut some small wedges and wedge the post in the hole. After you are done drilling, you can remove the press and will be able to put a C-clamp through that hole and clamp things from the middle of the table.

Creating an Airfoil Template

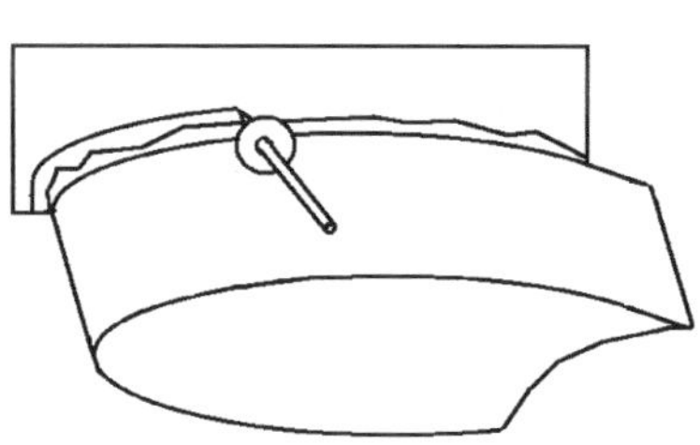

You can create an accurate template for the airfoil of your airplane with a few simple items. A template is useful to check the incidence of wing or tail for example. Take a sheet of flat material (plywood or cardboard) and cut out the approximate size and shape you want to create. Hold or attach the rough template to the airfoil. Find a washer or other round spacer with a pencil-sized hole in the center. Insert a pencil in the hole and roll the spacer over the surface along the template, drawing a line on it. This line is a copy of the airfoil contour, offset by the radius of the spacer. The template can then be trimmed to that line or be used to reverse trace it onto a different sheet.

To create a more accurate reproduction or "splash", prepare a board with roughly the right shape so it fits to the wing with gaps less than 1/4 inch anywhere. Cover the wing at the location you want the template at with clear tape to release it. Cover the template with Bondo and press it down on the wing on the released spot. Once the Bondo has cured, trim any excess off both sides. The template now has the exact shape of the wing in this location.

Wing Cradle

If you build your wings first, you have the advantage that they take up less storage space than a fuselage while you build the rest of the airplane. To store the wings vertically, you will need supports or a cradle. Two examples are shown here. To be able to move them around, mount them on small castoring wheels. Instead of cutting the airfoil shape from boards, which fit only the particular wing with that airfoil, you can make a sling support. The sling can be made from cloth or a piece of fire-hose.

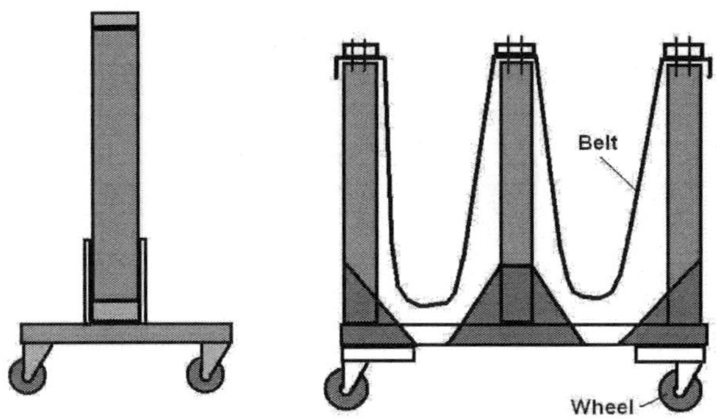

If you use cloth, select sturdy material and fold it several times into something that looks like a wide belt. Sew it lengthwise along the edges to keep it from unfolding. Nail or screw it to a stand like the one shown in the sketch. It will conform to any shape of wing which is narrow enough to fit between the posts. If the wing is too narrow, use pieces of foam to fill the space. You can either use two of these stands or one plus a sawhorse if the spars protrude far enough from the wing to rest them on it.

Storage of Control Surfaces

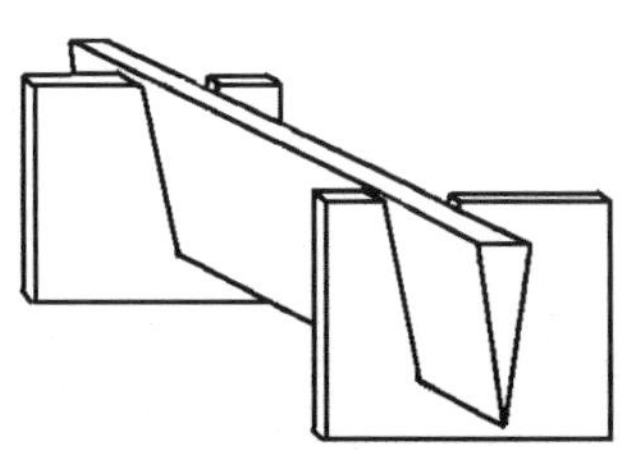

When it is necessary to work on a control surface trailing edge down, a good stand can be made from thick styrofoam plate. Cut a V-shaped notch in a rectangular piece. Use two of them to support the control surface. The styrofoam is soft enough to not damage the control surface. To build more durable stands, use wood and pad them with felt or foam. A piece of wood extending perpendicular to the board at the bottom of the slot helps support the trailing edge over a larger distance and reduces the pressure and the chance for damage. For storage of multiple parts, you can build a rack like the one shown on the left and hang it on a wall.

Straight Lines

It will often be necessary to build a component perfectly straight. For these cases you should keep a straight bar of appropriate length handy (4ft for example). It should be stiff enough not to deform easily. An example would be an aluminum profile of 1/8 inch wall thickness, either rectangular, L or T-shaped. It can then be clamped to or used as a guide for the part you are building to keep it straight. If you are just checking if something is straight, use a string. It needs to be pulled tight so it will not sag in the middle. Wire is not suitable because it bends easily and would require much more tension to keep it straight.

Breathing Protection during Spray Painting

A dust mask does not offer you any protection when you paint. The only kind of breathing device for spray primer like zinc chromate is a 3M 6000 series organic filtered respirator. If you buy a cheap respirator make sure it has the canisters installed for fumes and not the white filters for dust. The 3M respirators are not adequate for spraying the urethane aircraft paints either. One needs a re-breathing device or positive fresh-air supplied mask for this. This can be constructed from a full face mask, swimming pool vacuum hose and a small new vacuum cleaner.

Working with Composites

Riveting Thin Composite

Riveting composite is not a good idea, but sometimes it becomes necessary. Unfortunately, rivets will easily pull through thin laminate. To avoid the issue, bond a thin strip of aluminum to the inside of the areas where the rivets will be used. As an alternative, you can use a washer with the hole matching the rivet diameter on the inside of the composite to avoid pull-through and damage to the laminate.

Composite Skins and Metal Fasteners

You can see it on airplanes with composite cowlings and metal fasteners. Even if the fasteners are flush and painted over, the paint will crack at the fastener heads. There are several reasons for this and only one good way to avoid it. The fasteners are typically not clean enough for the paint to stick to them. If you degrease them before installing them it will help some. Under vibration, rivets can move slightly and any movement will crack the paint. It helps some to set them wet with adhesive. You can eliminate any cracking if you apply a thin layer of glass over the fasteners and finish and paint the composite with the rest of the skin.

Finishing Composite Airplanes

Preparation of the cured composite wing and fuselage skins for paint is a very time consuming task. There are different ways to do it. This one is a summary from a builder who used the Poly Fiber process.

Apply a thin coating of Poly Fiber Superfil over the whole exterior surface. This has the effect of 'sealing' the surface. The Superfil can be applied with a putty knife and heat gun. Then coat all surfaces with Poly Fiber Smooth Prime. Use open cell foam rollers. Make the Smooth Prime film as heavy as possible - 6 or more coats - by overcoating while it is still wet. Both products are epoxies, which shrink less than automotive (polyester) fillers and both are lighter than most other viable alternatives. There will still be some shrinkage, both of the structure and the primer, so it is a good idea to let the primed parts 'age' for a few months before sanding them. For final prep, take a long sanding block for leveling the surface. Before sanding the primer, spray a thin, dark 'guide' coat (flat black lacquer) to highlight low areas. It will be sanded off the high spots first, which makes the low spots stand out in dark color. Keep sanding on the high spots until all the dark paint is gone or fill the low spots. Time spent on getting the surfaces flat will be rewarded in the appearance of the topcoat. In many cases you can save money by doing the surface preparation for the final painting yourself and sending the top coat job to a competent automotive painter. Just make sure you tell him about any special considerations for your airplane if he is only familiar with cars. For example, paint can add a lot of weight if it is applied too thickly. Composite airplanes should avoid dark colors to minimize the chance for potential heat damage in the sun.

Cutting Fabric

If you are building anything from composite, you will need to cut fabric. If you are working with large pieces, you will need a cutting table. The table should be large enough that a roll of fabric fits on it without extending past the edges. It needs to have a smooth surface to keep the fabric from snagging on something. Keep the table clean of dust, grease or anything else which should not become part of a laminate. If necessary, you can cover a table with a heavy duty plastic foil to obtain a smooth surface. If you want to cut the fabric with a blade, the surface of the table will get cut as well. A plastic plate is the most suitable material for this job. If you are using scissors to cut the fabric, the table surface should not get damaged.

Whether the fabric is glass, carbon or Kevlar, the scissors needs to be kept sharp. Some scissors work better than others, but in general they should be sturdy. Even cheap ones may cut very well.

Composite Access Door

You need an access door in your composite wing skin? No big problem. You can make it any shape and get a perfect fit even on complex curves. If the door does not need rounded corners, the simplest method is to cut the door directly from the skin with a thin rotary or vibrating blade. In most areas it is better to round the corners of the hole to avoid stress concentrations. In this case a different method is easier to use. The only requirement is that you have to have a way to reach the back of the door area to build up a backing lip for the door. Pick the area where you want your access door and sketch out its outline on the surface of the basic piece. Then cover this area with packing tape (or any smooth tape from which a layup will release) a few inches past the boundary of the door area. Now apply the door wet layup to this area. I would recommend using at least the same number of plies as the original skin has in that area. After it is cured, mark its position so that you can get it back in the same place after you have trimmed the door hole to the exact shape you want. Put this door blank back in the position where you want the access hole. From the skin backside, using an ultra fine point Sharpie, trace the outline on to the skin by going around the cutout. Cut the hole leaving just a bit more material than the Sharpie line on and sand down progressively all the way around the door until it can be lightly force-fit into its hole. Now you can tape the door down flush with the outside contour of the basic skin, cover it with packing tape and apply an appropriate lip on the basic skin via wet layups. When this is done and the lip is cleaned up to its final shape, match drill through the door and lip to position screws / nutplates. Install nutplates and screws. Take the door off and go around the edge with light sandpaper until you have the gap you desire between the door and the skin. This should be two times the paint thickness and then a bit more, as desired. If you follow this procedure, the door will come out perfectly with no mess or micro/flox sculpturing. And your contoured doors with perfect fits will look good.

Resin and Hardener Dispersal

Resin and hardener are mixed at a certain ratio which depends on the type of resin. A common mix ratio is 100:25 or 100:22. To mix resin using a scale, use the mix ratio by weight. If you are using pumps, the mix ratio is by volume, which is slightly different from the mix ratio by weight. Pumps are available from some manufacturers but are only valid for one mix ratio. Instead of spending money on those pumps, you can use soap dispenser pumps. They come as part of liquid soap bottles. Use up the soap, clean out the bottles and fill one with resin and one with hardener. Make sure they have the same capacity per stroke (typically 2 gram). To obtain the correct mix ratio by volume for a 100:25 (= 4:1) resin system, use 4 strokes of resin plus one stroke of hardener. For a mix ratio of 100:20 (= 5:1) use 5 strokes of resin and one stroke of hardener. Beware that the hardener is corrosive, it will eat through a plastic bottle eventually. A soap bottle with hardener can at the most be used for a few months. Long-term storage for hardener should be in its original container or a glass or metal container.

Containers for Mixing Resin

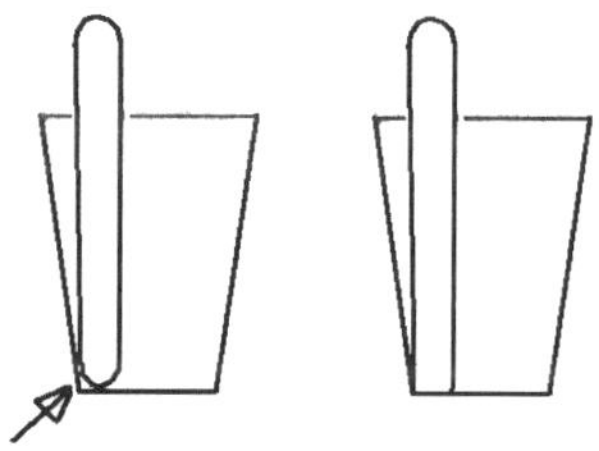

Ideal containers are paper cups, which can be bought in many stores in various sizes. You can also save and clean cans or plastic cups from the kitchen. They should have smooth sides, otherwise unmixed resin can remain in the grooves. Dentists' spatulas make good mixing sticks. Usually they have rounded edges. To avoid missing unmixed resin in the corners of a container, trim the mixing stick corners square. After use, let the resin cure on the stick, sand it smooth on a belt sander and reuse it.

Resin Brushes

You could clean a resin brush after use, but this requires a lot of solvent and takes time. Some of the resin will remain and cure and shorten the usable bristles every time you used it. It is easier to partly fill a glass container with solvent and drop the brush into it. It has to have a lid which closes tightly. The brush will stay soft and is ready for use the next time you do a layup.

Working with Narrow Strips of Fabric

Narrow strips of reinforcing fabric (glass or carbon) can be tricky to handle. They distort, separate, fall apart and stretch just to name a few problems. Rather than applying them ply by ply, consider pre-wetting them on the bench on a sheet of plastic and laying them in place together. If you use peel ply as the outer ply, it makes picking them up from the plastic a lot easier and prevents distortion.

Layup at Lower Temperatures

Composite layups should be done in a warm room to allow the resin to flow well enough to completely wet the cloth. If the temperature is too cold, dry spots may remain in the cloth and the resin may not cure properly. If it cannot be avoided to do a small layup in cold weather, use a hair dryer or heat lamp after the resin was spread to warm it up. This reduces the viscosity and allows it to flow. Even though the resin will cure to a certain degree at the low temperature, for the laminate to reach its full strength, it needs to be post-cured at higher temperatures. To fully cure the layup, build an enclosure around it. This can be a simple as a cardboard box taped over the layup area. Use an electric heater with a fan to duct the hot air into the box. Monitor the temperature in the box with a thermometer.

Keeping Resin Warm

If your shop is rather cool, you may want to keep the resin warm and ready for layup separately rather than heating the shop all the time. A way to accomplish this easily is a heater box. This consists of a insulated box with a door and a light bulb rigged inside. A switch on the outside lets you turn the lightbulb on or off. Place the resin and hardener containers inside the box and turn the light on. A 100W lightbulb will keep the temperature in the box up and your resin liquid. When you build the box (plywood is a suitable material), make sure that the insulation is not touching the lightbulb, so it cannot set it on fire. If you plan on doing a layup in the morning, you can turn the light on in the box the evening before and have the resin warm and ready for use the next day. The box should be large enough that the amount of resin you typically use per day fits in it, but not so large that a single lightbulb could not warm it anymore. For example, a good size is 1 ft x 1 ft x 1 ft.

Spreading Resin

Select the most appropriate tool depending on the size of the layup. This may be a brush, a roller, a resin mixing stick, a squeegee or spreader. For a large, flat surface, pour resin on the cloth and spread it with a squeegee. Excess resin can be pushed out towards to the edges. Layups can often be completed quicker if the cloth is pre-wetted on a flat surface (table). Invert the sequence of the layup: a sheet of plastic, peel ply, layers of cloth as needed. The whole layup can then be placed in the mold as an assembly. This works best if the contour is not complex.

Carbon Fiber and Corrosion

Carbon fiber is conductive. If a laminate made from this material comes in contact with metal, especially aluminum or carbon steel, the metal will corrode unless it is well protected. It is best to avoid using metals which easily corrode in combination with carbon. Aluminum is particularly vulnerable. Use stainless steel, brass or bronze wherever you can. Where that is not practical, place a layer of glass between the carbon and the metal. With an Ohm-meter, check that there is no conductivity between the parts and the carbon after they are in place.

Composite Tubes

Composite tubes can be used for many jobs: as bushings, conduits, stiffeners, or anything else you would need a tube for. Composite tubes are lighter than metal, do not corrode and can be bonded to other assemblies.

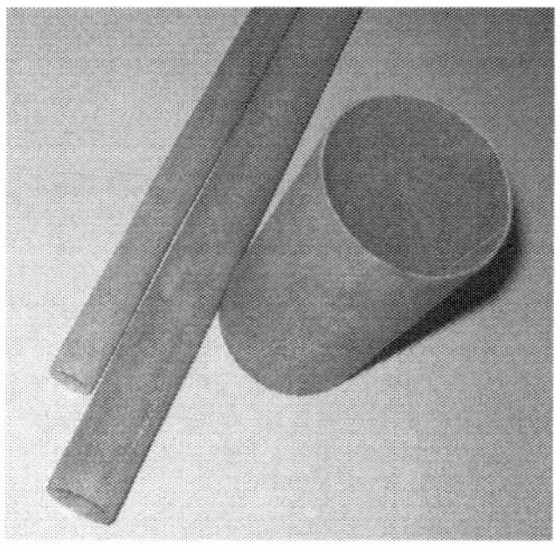

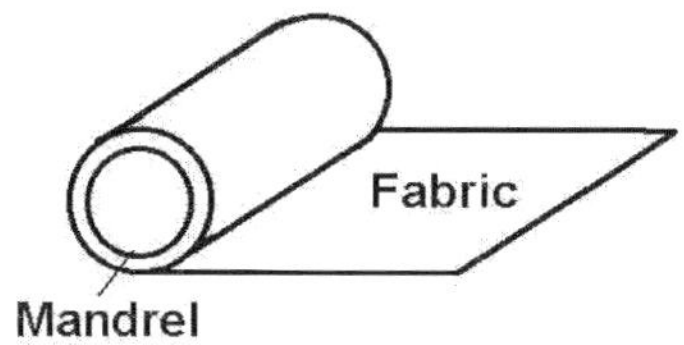

To make a composite tube, select a mandrel of the shape and diameter you need. It does not have to be round, but must not have sharp corners. The outside diameter of the mandrel will become the inside diameter of your tube. The wall thickness depends on how many plies you use. Roll a piece of fabric once around the tube and measure its length. To get a wall thickness of 6 plies, cut a piece of fabric six times as long as the one you measured. Instead, you can use two plies together and roll them up three times. It is easier to use fabric with 45 fiber orientation because it is less likely to develop wrinkles as it is rolled up. But you can also use 0 /90 or unidirectional fabric or a combination of all. To start, wrap the tube with one layer of thin plastic sheet to keep the laminate from sticking to the tube. Do not tape the plastic to the tube. It should slide easily over the tube so the layup can be removed after it is cured. Find a way to support the mandrel horizontally at the ends. You need to be able to rotate it for the layup. Prepare the fabric on a sheet of plastic for wetting it out with resin. Then wrap the fabric around the mandrel. It takes a bit of practice to get it on without wrinkles or distortion. You can even taper the wall thickness by changing the shape of the fabric from rectangular to triangular. To improve the exterior of the tube, you can wrap peel ply as the last layer around it. Another method for surface finish is shrink tubing over the layup. When the shrink tubing is heated, starting in the center and working out towards the ends, it squeezes the excess resin out and the surface will be smoother than without it. After it is cured, remove the mandrel. This should be easy as long as the protective plastic did not stick to it. The length and diameter of the tubes fabricated by this method is only limited by the size of what you can physically handle. Small ones (less than ½" / 12 mm diameter) are more difficult to make than larger ones.

Fillets

Most layups require fillets of filler material like a paste of cotton flox or micro in corners. Small fillets can be applied with a brush if the material is not too stiff. Round the corner of a mixing stick to the shape of your fillet and use it to scrape the edge. Drawing your finger along the edge works well for small fillets. Adding a little Aerosil allows the paste to flow smoother.

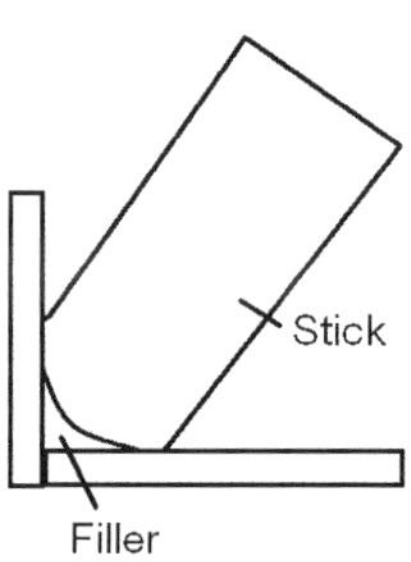

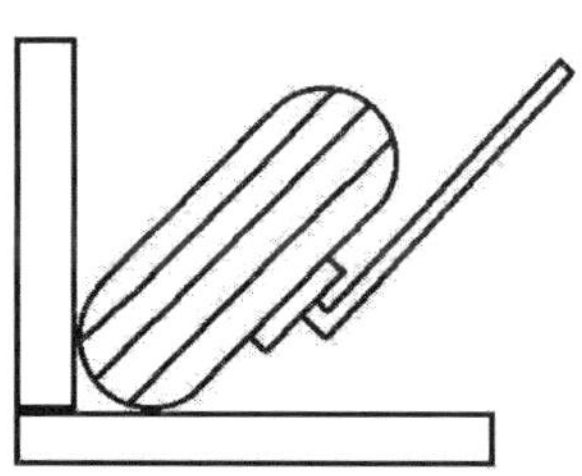

Larger fillets which are made and cured separately from a layup can be shaped with a roller. A model airplane tire works well for this job. Here is a description of the technique. Chose a tire that matches the desired radius of the fillet. Manipulate the fillet material roughly into the corner. Put peel ply over it and work it down into a fillet, first with your fingers, then with the roller. After several passes with the roller, you should get the shape you wanted. Excess fillet material may start to migrate out the sides. Clean it up before it cures.

Larger fillets are not necessarily better fillets. For example, for fillets on a composite wing rib, the purpose is to provide a support for the gusset between the rib and wing to stick to so that it does not bridge in the air. In this case, the smallest fillet that does the job and keeps the fiberglass gusset from bridging is the best fillet because it adds the least amount of weight.

Composite Landing Gear Leg Root Fairings

This is a description how you can make your own landing gear leg root fairings from composite. It requires some sculpting skills. Obtain some suitable foam like Styrofoam from a local hobby supply store or builder's store. The foam is either used to make a temporary mold and in this case will not be part of the finished part or it will be used in the part to stiffen it. The second method is described here. Cut the blocks at the same angle of the gear attachment to the underside of the aircraft. You can glue blocks together to get to the required thickness. Cut slots in the foam and fit them under the fuselage and around the upper portion of the landing gear. Here is a picture of the foam blocks and the finished shape.

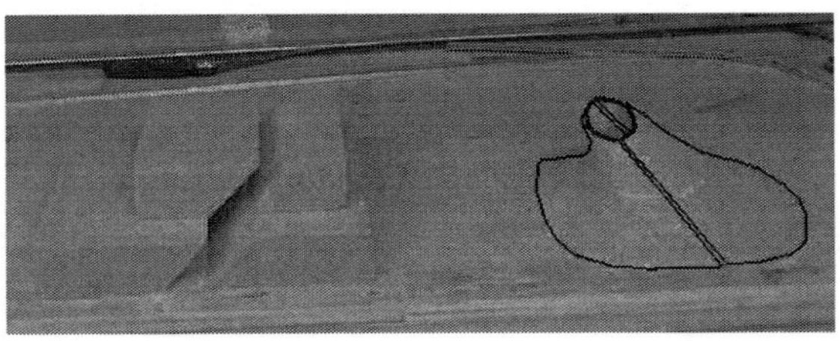

Once the fit around the gear is obtained, the rest is just a matter of sanding the blocks to the desired shape. Sanding the foam is no problem but keeping it from disintegrating at the edges and not breaking it as it gets thin can be a bit interesting. The first step to laying glass over the form is to fill the open cells of the foam with a micro-epoxy mix. You can let the mix cure and sand the form with 60 grit sand paper to prepare it for the glass covering. If the foam is not going to be part of the finished part, cover it with tape to allow it to release. Because of the complex curved shape, this will require a lot of small pieces of tape. An alternative is to lay up directly on the foam, allowing the layup to stick to it. Afterwards the foam can be broken out and the inside sanded smooth.

The glass is laid up in multiple small overlapping fabric pieces because the shape is too complicated to do it with one piece. Multiple pieces make it easier to follow the form. After application of the glass plies, strips of peel ply are added to soak up excess resin and to ensure the finished product is smooth and requires less sanding to finish. After the layup is cured, trim the edges. If you only used the foam as a mold, remove it. You probably still need to do some filling and sanding to get the final shape right. For easier installation, cut it open on one side.

Poor Man's Vacuum Bagging

Wet layups without core can usually be done without vacuum bagging, which is a way to apply pressure to a laminate. Vacuum bagging helps squeeze out excessive resin, presses the laminate against the mold and reduces the chance of voids. Of course not every one has the equipment needed for vacuum bagging. If you have an application where you need some fairly uniform pressure on a laminate, sand is a simple alternative. It can be filled in a bag and placed on the part or the part can be covered with thin plastic foil and sand is dumped right on top of it.

These pictures show the fresh laminate, in this case an armrest in a door frame, covered with plastic on the left and everything covered with sand on the right. After it is cured, the sand is removed, the plastic foil pulled off and the part is demolded. The picture in the center below shows how foam stiffeners covered by a carbon ply are bonded to a skin with sandbags applying the necessary pressure. You could also use Ziplock bags filled with sand for smaller parts. Do not overfill the bag to allow it to be flexible enough to "flow" around the part.

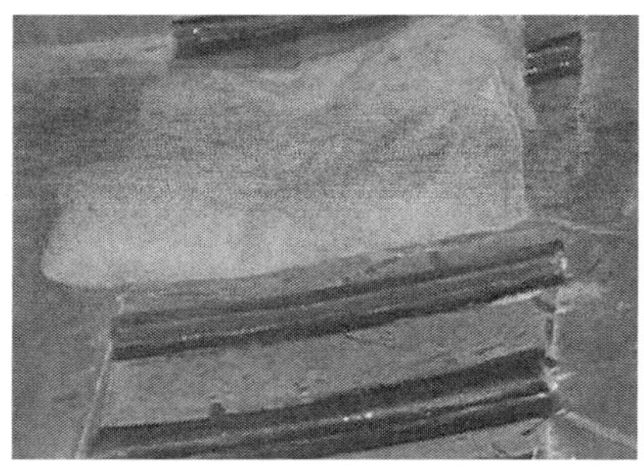

Usually vacuum bagging is required to achieve a reliable bond between face-sheets and core if sandwich construction is used. A more labor-intensive alternative if foam is used as the core is to fill the porous foam surface with microballon prior to the layup of a face-sheet. You will need to use a lot of excess resin to make this work, because the foam soaks it up and the fabric may not become sufficiently wetted. This picture shows a layup of carbon fiber on foam. Use a squeegee to spread the micro on the foam and also to spread the resin on the fabric. Special care is necessary to prevent air getting trapped between the foam and the fabric. Especially with carbon fiber, air bubbles under the fabric are hard to see. Light reflecting off the surface can show a distortion if a bubble is trapped. Keep pushing with the squeegee until all the air has been worked out or it will show up later as bumps requiring a repair. Really persistent bubbles can be brought under control by pricking them with a sharp object (needle, nail) to let the air escape.

Reducing Gaps

Access panels or cowlings with excessive gaps just do not look good. This is an easy way to produce thin, clean gaps: Assemble the parts and spread microballon on the gap as thin as possible, do not use release tape. When the micro is slightly firm, but not yet set up, take a knife blade and run it around the joint. It is easy to stay in the joggle, just keep the knife blade against the cowling. Let it cure, sand smooth and remove the panel or cowling and you end up with a great seam.

Bonding Nutplates

The threads in nutplates are sometimes too tight to use them for Philips head screws and a screw driver, especially the smaller sizes (10-32 or smaller). This is the result of the way the threaded part is deformed to make it lock-nut. Look at it, it is slightly oval. Before bonding the nutplate on, put the oval part in a vise and gently squeeze it back towards round. That will make it easier to install the screw.

There are basically three ways to attach nutplates to composite structure. The first method is to rivet them conventionally, no different from metal structure.

The second method is to bond them (not for floating nutplates). First, drill the hole for the bolt. Prepare the nutplate by sanding the bottom surface of the nutplate to roughen it. Sand the surface the nutplate will be bonded to. Clean off any grease by dipping the nutplate into a container with solvent (alcohol, acetone). Prepare the bolt or cleco which will be used to hold the nutplate in place by waxing the threads. Mix resin and cotton flox and wet the surface where you want to bond the nutplate. Insert the bolt through the hole and into the nutplate to hold it in place. Thread it in far enough so that it will keep the resin out of the threads. It does not need to be holding the nutplate tight. That can be done with a small clamp or clothespin on the portion of the bolt sticking out. Then spread the cotton flox over the nutplate until it is covered completely. Once it is cured, the bolt should come out easily.

The third method is a combination of the first and the second method. Rivet the nutplate to a piece of scrap composite material. Then bond the scrap piece with the nutplate to its intended location.

Bonding Aluminum

Bonding aluminum for primary structure is out of the reach for the average homebuilder. It requires special surface preparation (etching) to ensure strong, reliable bonds. But it is certainly possible to bond secondary structure and small parts such as brackets with good results. Keep in mind though that your bonds may or may not be as good as expected. Besides sanding and etching a part, drill holes in the part which will be filled with adhesive. This helps lock the part into place. If possible, add a few plies on top of the aluminum to further secure it. An example is the attachment brackets I used for my tail wheel installation.

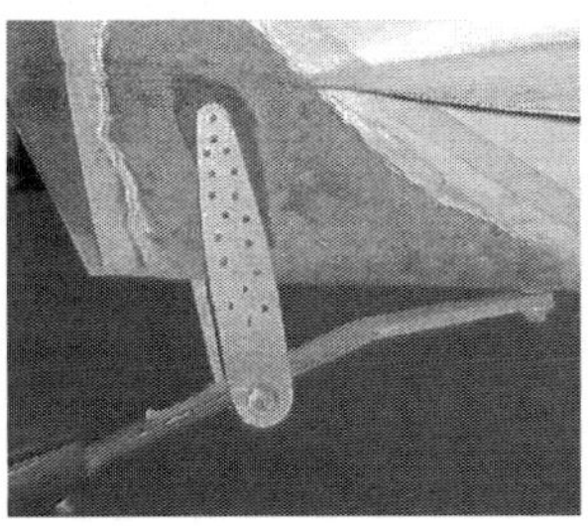

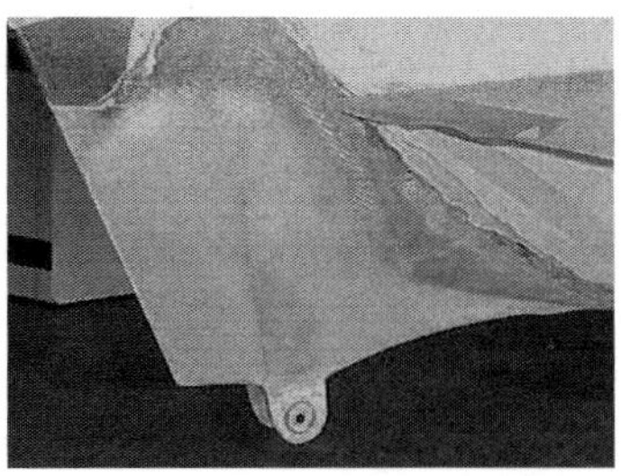

Tailspring brackets ready for bonding (left) and glassed over, bushings bonded in (right).

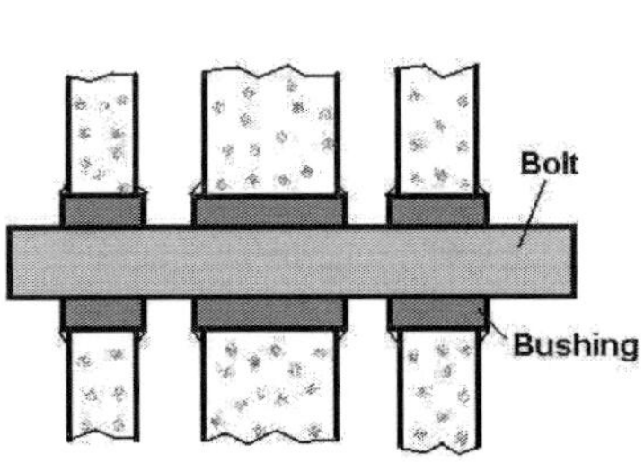

Bonding Bushings

If you need to bond bushings in several parts which have to be perfectly aligned with each other, a bolt should be inserted to hold them aligned. The problem is, when you have inserted the bushings in their holes, nicely coated with cotton flox, and try to push the bolt through, you are more likely to push the bushings out than the bolt in. The solution is to let the resin set a bit until it is gelling; then you can push the bolt in without the bushings sliding out. It means you have to wait and keep an eye on the viscosity of the resin until it is just right. The resin must not be too far cured, the bushings should still be able to move slightly for the alignment. To speed the process up, use heat from a hair dryer.

Inserting Tubes Through Ribs

You have a wing with many ribs and access to them because the wing skin is not yet in place. You want to run a tube through those ribs and need to cut hole the size of the outer diameter of the tube in precisely the right place in each rib, lined up straight. This could for example be a tube for control surface actuation or a long capacitive fuel probe introduced for better fuel level reporting.

Typically the hole through the inside rib (nearest the wing root) is easy to define. And the end point of the tube needs to be defined, even if it is a point in space inside the wing. But how do you drill the holes in the ribs in-between? Of course, there is trial and error but this is to be avoided as it does not work that well. We cannot stretch a string directly between the start and end points, but we can use the concept of a parallel line. Stretch a string above the wing that is just above and parallel to the line defined by the start and end points.

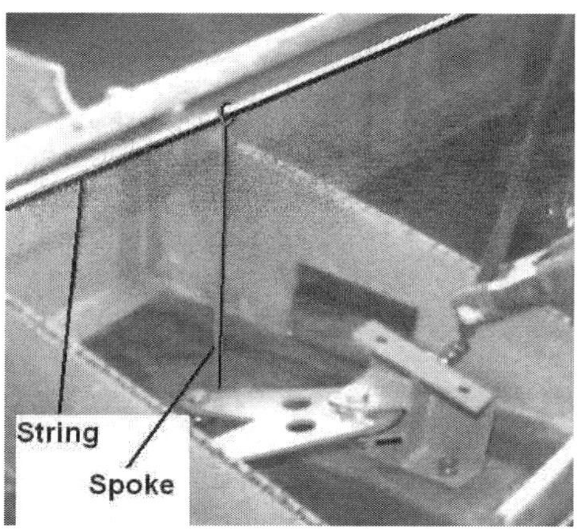

Vertically the string should be as close to the top of the wing as possible at the starting point. The end point will depend on how your tube needs to be located. Now take a straight piece of wire (a bicycle spoke works great) and make a loop on one side and put it on the string. This will act as a mini plumb-bob. Cut the end of the wire so that its end matches the start point. Move the wire to the end point. Position the string vertically and horizontally so that the end of the wire points exactly to where the end point of the tube will need to be. Clamp the string so that it is held tightly in that position. The wire can now be swung up and around each rib. So move the wire, rib by rib, let any movement stop and mark the end of the wire on each rib. Simply drill at the marks and the tube should insert almost perfectly. If you have a thick composite rib, you can angle the drill slightly to match the angle of the tube. In the picture above you can see the string and bicycle spoke wire (left) and the end result with the tubes installed (right).

Clay as Tooling Material

Clay is a wonderful material to shape complex parts with. I have used it many times when I wanted to build a single small part of a shape I did not have a mold for. Just sculpt the part and cover it with a few plies. Once it is cured, scrape the clay out. I made tail wheel fairings and reshaped my cowling inlets with this method.

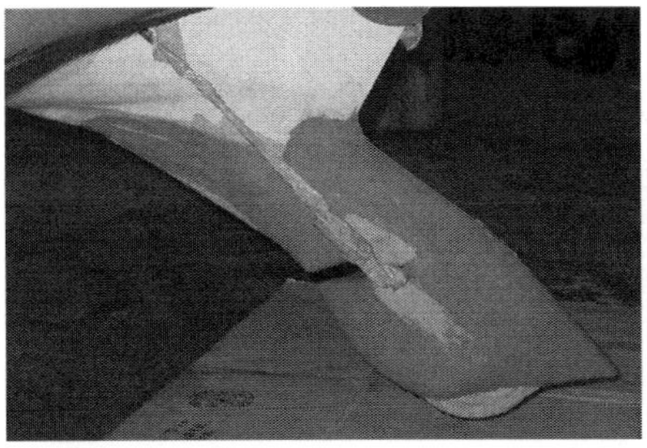

Tailwheel Fairing

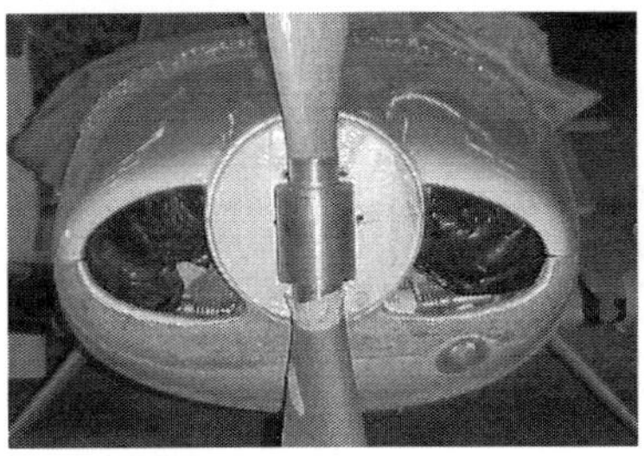
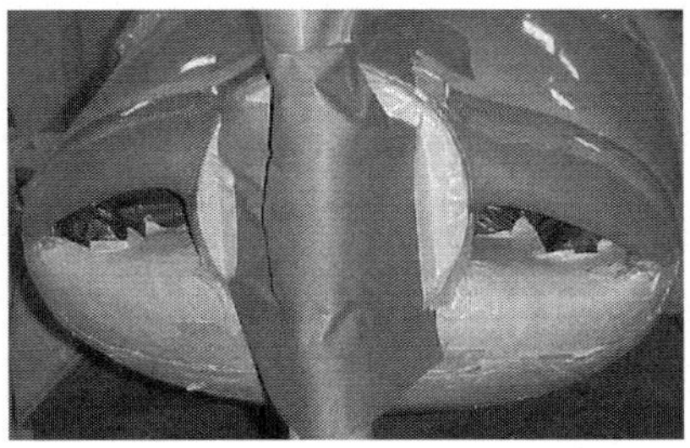
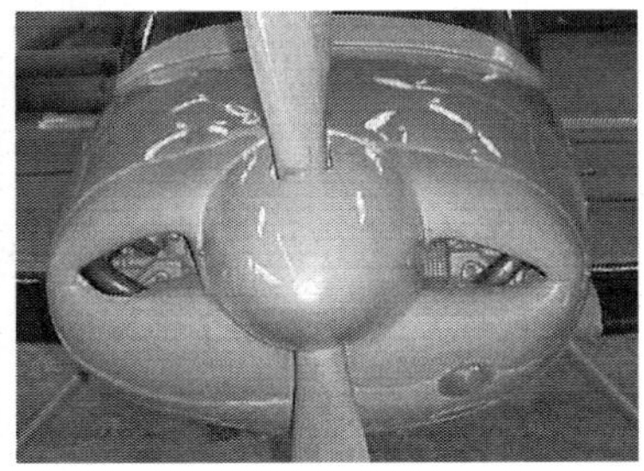

Cowling

Control Surface Actuator Horn Fairing

The actuation horns of control surfaces often stick out into the air stream (left picture, no fairing). This causes drag and exposes the bearings to the direct effects of dust, rain and other debris. You can easily lay up a simple fairing which will lower the drag and protect the bearing. Take a piece of card stock or thick paper and cover one side with tape to release it. Trim the piece into sort of a cone shape, this will take some trial and error to get a good fit. Bend it to a shape like the one shown in the right picture to see if it covers the hole and provides sufficient clearance for the fitting with the bolt. Once you are satisfied with the shape, prepare the plies for the layup (3 plies work well). They should cover the cone and provide a one inch flange past the cone all around. Pre-wet the plies, place them on your cone (flat) and place this assembly on the surface you have prepared for bonding. When you do this, bend the cone into the shape you want. Hold it in place with small weights placed around the edges of the cone while it cures. Do not place any weight on the cone, that would distort the shape. Once it is cured, remove the cardboard and trim the edges.

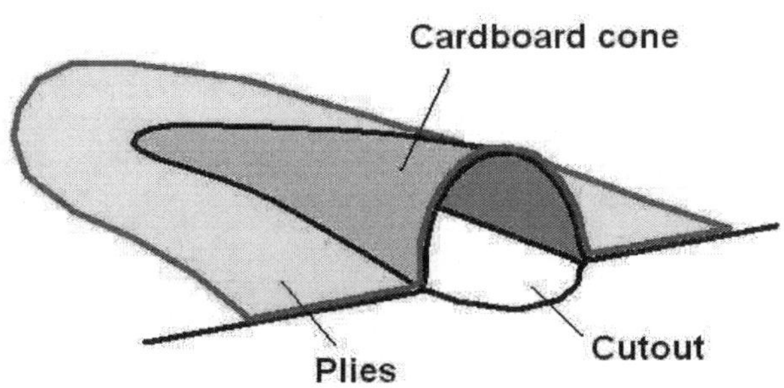

Sanding Wings

To obtain a smooth surface without waves and low spots, glue sandpaper to the bottom of a long, flat sanding block. Sand diagonally in both directions. The high spots will be sanded down first. To make the low spots more visible, apply a light coat of spray paint of a different color. This will be sanded off in the high spots and remain in the low spots. Alternatively you can hatch the area to be sanded with a pencil or mix a little bit of paint with a lot of acetone and wipe it over the surface. If the surface is concave, use a contoured sanding block.

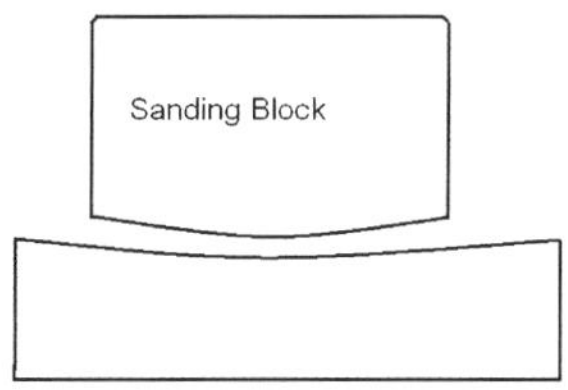

Finishing the Bottom of the Fuselage

If your airplane is going to be large or heavy and you do not have a jig to rotate the fuselage to different angles, consider doing some of the prep, prime and finish work on the fuselage bottom side before the shell (halves) are bonded and become hard to handle. If you don't like to sand something above your head, with the dust drifting down on you, this can help you minimize it. Prime and sand the bottom side of fuselage at this stage, it will be easier to reach instead of sanding and painting upside down. Do not apply primer to areas to be bonded.

Drilling Composites

It is difficult to drill holes in composite material without causing some delamination and fraying around the edge of the hole. The most common cause is too much pressure in combination with a dull drill bit. The strength of the resin keeps the fibers together, but resin is much softer compared to metal. When the drill bit exits the back side of the laminate, the pressure tends to separate fibers from the laminate, causing the edge of the hole look splintered. A dull drill bit tears rather than cuts the fibers and causes delamination of the entire hole's wall. Drilling heats up the material which further softens the resin. Composite material has poor bearing strength to start with (1/3 to ½ of aluminum depending on fiber orientation, material and resin content). Delamination will further decrease the bearing strength. Observe the following to minimize delamination:

- Use a good backing plate.
  If the same backing plate has already been used to drill 20 holes, it will have more holes than smooth surface. Use one without holes. Wood may be too soft.

- Use a sharp drill bit.
  A regular carbon steel drill bit will become dull after drilling just 2 or 3 holes through glass or carbon. It is slightly better to use carbide drill bits, but even they need to be sharpened frequently.

- For holes much larger than 0.25 inch (6 mm), diamond-coated hole saws give the best results. They grind rather than cut.

- A tapered drill reamer has three straight flutes and will cause the least amount of fraying on the backside. The first half-inch of the bit is gently tapered, with a sharper taper right at the tip. This bit cuts very cleanly with little heat generated.

- Dagger bits are solid carbide. They are basically flat with two cutting edges and cut nice holes. Resistance is a bit less than the drill reamers while heat generation is comparable to tapered drill reamers. The back side of the hole shows the biggest difference. Fraying is much worse and may even extend beyond the edge of the hole. Backing is definitely required with the dagger bits.

- Countersinking
  If you drill the hole from the opposite side of where the countersinking will be, it will remove the frayed material. Unless you use a carbide countersink, it will dull very quickly. The part needs to be firmly clamped or the bit will chatter.

- Minimize the Pressure.
  Use the least amount of pressure on the drill that will do the job. Use slow speed to minimize the amount of heat generated. Use an electric drill rather than an air drill to better control the speed.

- Drill the hole undersized and use a file or reamer to open it up.
  A diamond or carbide coated file can be used to carefully ream a hole to remove small amounts of delamination. For larger holes, use a tube slightly smaller than the diameter of the hole wrapped with a piece of sandpaper.

- Repair
  If small delaminations have happened, they can be repaired with a small amount of resin. Use a fastener of appropriate size for the hole. Release the fastener with wax. Coat the inside of the hole and the fastener with resin and insert it. Once the resin is cured, the fastener can be pulled out. It will be a tight fit which can be loosened up by slightly sanding the hole. Larger delaminations require replacement of the damaged plies and should be performed like any structural repair (see "Composite Facts" by Sonja Englert for details on composite repair).

Composite Fittings

Fittings can be made lighter and from non-conductive material if you use fiberglass rather than steel. They are easy to make and not expensive. The version described below will hold about 200 lb before breaking and can be custom-made to any size you might need it.

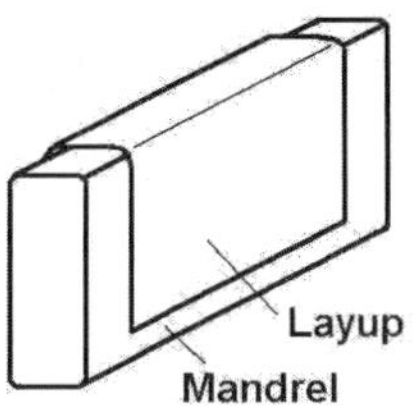

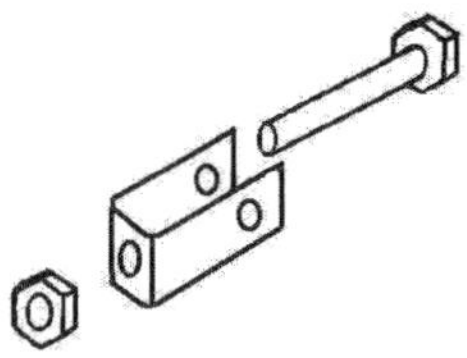

You need a suitable mandrel to do the layup of the U-shaped part of the fitting. Its outside dimension will be the inner width of the fork. The width needs to be slightly wider than the head of the bolt which will be bonded in later. The corners need to have a small radius (3/32 inch/ 2 mm). Prepare the mandrel by waxing it or covering it with a layer of tape. Prepare the layup on a flat sheet of plastic for prewetting. The first ply should be peel ply, then at least ten plies of 9 oz/sqyd fabric, alternating 0 /90 /45 fiber orientations for good bearing strength. As the last ply again use peel ply. Fold the layup over the mandrel as shown in the sketch. To obtain a flat surface and to squeeze excess resin out, clamp two flat plates to the sides. When the layup is cured, remove it from the mandrel and cut off pieces of the width you want the fittings to be. Drill a hole in the base for the bolt to fit through (1/4 inch bolt, fully threaded). This will be bonded to the fitting with a nut at the base to hold it in place. I used stainless steel bolts and nuts. They need to be etched prior to bonding with a solution suitable for stainless steel. Bond both bolt and nut with cotton flox. Drill the holes in the fork. To improve the bearing strength, bond washers to the outside of the holes in the fork or bushings to the inside.

Hardpoints and Bolts in Sandwich Structure

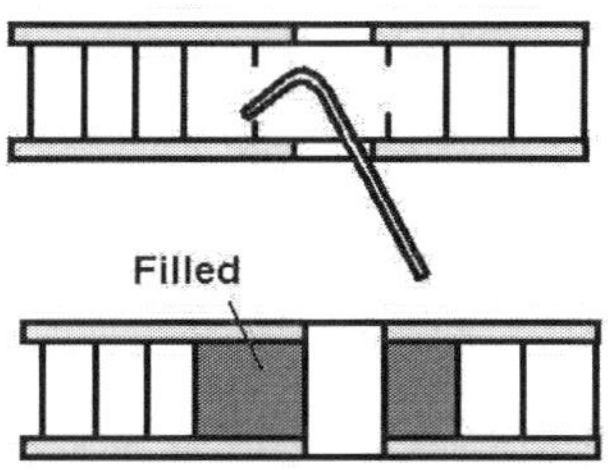

If there is a need to have a bolted joint in a sandwich panel, you need an insert to keep the core from being crushed. The core of a sandwich, whether it is foam or honeycomb, is too soft to tighten a bolt against. If you know ahead of time where the fasteners will go when you do the layup of the sandwich, you can build hardpoints into the structure replacing the core locally. Suitable materials are solid laminate, phenolic plate, even wood or aluminum if necessary. If it is a small area, you can just remove the core and fill it with cotton flox. This can be done even after the sandwich has already been cured. Drill a hole where the bolt will be used. Bend a piece of wire, insert it through the hole and scrape the core out in an area of at least twice the bolt diameter. Fill the space with cotton flox and let it cure. Now you can drill the final hole and tighten the fastener without the risk of crushing the core.

Making a Mold for a Wheel Pant

Sometimes it is useful to have a mold for a part you are building. Once you have a mold, you can fabricate multiple identical parts in case you need a replacement. Wheel pants are a good example because they easily get damaged. I will use making a wheel pant mold as an example of the mold fabrication process, but the procedure is applicable to all molds. First you need a plug. The plug is the part you will be duplicating exactly. The plug needs to be robust and stiff enough to survive the mold fabrication process. If the plug is too flexible, you risk making a distorted version of what you intended. In the case of the wheel pant you can use the original wheel pant as the plug, but it may need to be reinforced with a frame. Or you can fabricate a plug from any other suitable material like wood, foam or plaster.

Once you have your plug, make sure that any small defects like voids, scratches or dents have been filled and smoothed, because everything will be reproduced exactly in the mold as it was on the part. You might as well spend the work up front and make the plug as perfect as possible rather than having to rework each part you pull from the mold. The plug's surface should be wet sanded with 400 grit sandpaper or higher and the finer, the better. It is possible but not necessary to polish it. Any features which you want to transfer from the part or plug to the mold like trim lines or the location of the axle must be present on the plug. If you want to have a hole location marked on the part, use a drill bit to just work a small depression into the plug. In the mold it will stand out as a dimple and on the final part it again will be a hole.

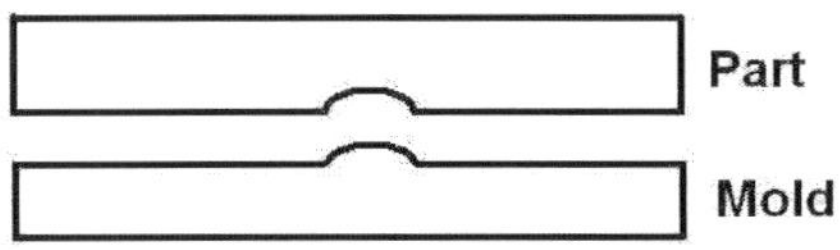

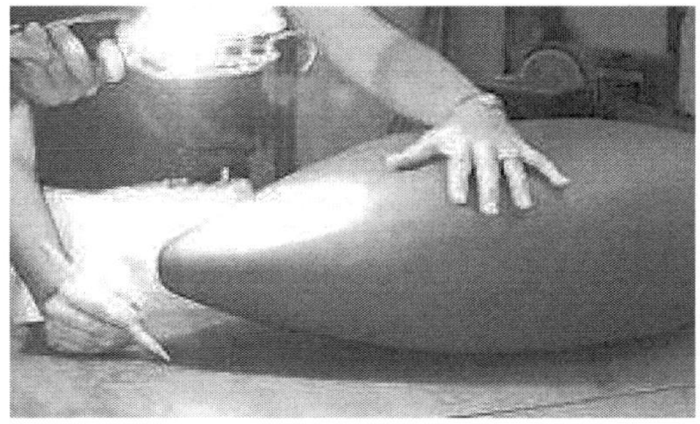

In the case of the wheel pant, the mold will be made in two halves to be able to demold the part. Take a sheet of Masonite or something similar and cut out the shape of the wheel pant so that the pant will fit into the hole. To mark up the shape of the part as closely as possible on the board, shine a light on the edges of the part and trace the outline of the shadow on the board. This is not totally precise, but suitable for our purpose. You can use a saber saw to cut just inside of the traced line. Cut and try until you get a good fit to just clear the plug. The plug will be set halfway into the hole. It is important to have the parting line exactly at the thickest part of the plug, otherwise you would create an undercut which would make it impossible to demold the part. Mark the centerline on the plug.

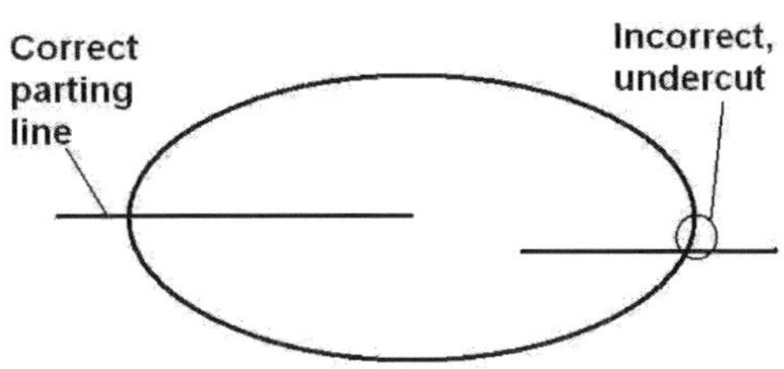

Use a box to support the plug and Masonite board so it does not shift during the layup. The box must be deeper than half the thickness of the plug. What you see sticking out will be used to make the first half of the mold.

Fill the small gap between the board and the plug with modeling clay. The more precise your fit was, the less clay you will need. If the gap is too large, more than 1/4 inch, you need to add scraps of wood underneath to keep the clay from pushing through the gap.

Use a stick to push the clay down into the gap. Once the gap is completely filled all around, scrape the excess clay off with the stick so that you have a nice sharp corner. Once that is done, apply at least three coats of wax to the surface of the plug, the clay and the board. Use mold release wax specifically designed for this purpose. Let the first coat of wax dry for about an hour and gently buff it off before applying the next coat. Do not rely on the wax alone to release the part. A second release method is PVA (Poly Vinyl Acetate), a liquid agent in an alcohol base. It is applied by a spray unit to cover the surface in three or four mist coats. When it dries, it forms a film

like a thin sheet of plastic. Allow the PVA to dry at least 10 minutes between coats and wait at least an hour after applying the last coat before proceeding. The PVA must be completely dry. Warming the PVA up will help making the mist finer. You can keep the PVA container in a bowl with warm water. When you apply it, you need to be further away from the part than you typically would be when spraying paint. All you need is a fine mist. With both the wax and the PVA you should be able to get the mold removed from the plug without too much trouble. This completes the preparation of the plug for the actual mold fabrication. Only start the next process if you have enough time to finish it, because you must not interrupt it.

For a smooth and durable mold surface, the first coat is a surface resin. This is a very viscous, pastelike material which is brushed onto the plug. Try to apply two coats of it evenly on the plug and what will be the flange. The flange should be at least two inches wide. The mold surface coat is an epoxy with filler which is available in different colors. The filler increases the wear resistance of the mold surface. Allow the surface resin to cure until it is no longer sticky, but not completely hard.

If you touch it with a stick, it should leave an impression but not stick to it. Now you can proceed with the layup of the mold plies. Before the first ply, use cotton flox mixed with resin to form a fillet in the corner between the plug and the board. It should be at least 1/4 inch or larger to keep the cloth from bridging the gap and cause a void in that sharp corner. Coat the surface resin with laminating resin and lay the first ply over it. To make the mold sufficiently stiff, you will need 10 to 15 plies of 9 oz/yd^2 glass cloth or less plies if you use thicker material. It does not really matter what resin or cloth you use as long as it meets the requirements for thickness, stiffness and temperature resistance. If the mold is used for vacuum bagging, it must also be sealed well enough to hold the vacuum. Laying up or applying that many plies take quite some time. Keep your work area warm enough during the layup to allow the resin to cure.

Once the layup is cured, you need to remove the board to be able to make the second half of the mold. Leave the plug in the mold you just layed up, just trim the flanges to be slightly larger than what you want to end up with. Release the plug and the flange from the first half of the mold like the first time with wax and PVA. Repeat the steps of the layup and cure the part. Do not try to demold it unless it is fully cured. Trim the flange to the final size and drill 3/16 inch or 1/4 inch holes in the flange all around, spaced 6-8 inches apart. This way the two mold halves are still perfectly aligned. Bolts will be inserted in these holes to hold the two halves together and aligned when you make a part in the mold.

 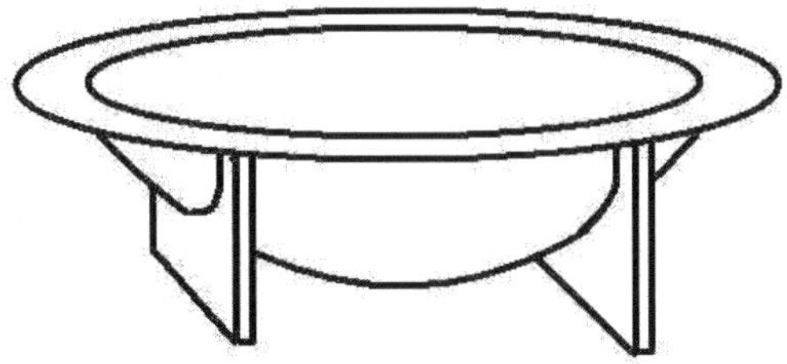

Gently pry the mold two halves apart. A metal tool would cause scratches, use wooden spatulas and sticks for this job. To remove the PVA film, soak it in water. If the surface quality of your plug was good, you should have a good mold surface. If not, it can be repaired with the surface resin and sanded and polished. Chances are that the mold of a wheel pant is rounded and will not sit well on a table for the layup of a part in it. It is convenient to add ribs which will serve as a support for the mold during the layup. These ribs can be cut from wood and bonded to the mold with flox and a few plies of fiberglass.

Working with Metal

Cutting Straight Edges from Large Sheets

For cutting sheet metal, offset blade hand shears are very useful. They are harder to find than the regular aircraft shears but worth looking for. If the sheet metal (aluminum) is .032 inch or less, and you have a nice narrow straight line drawn with a Sharpie marker, it is possible to cut very straight and with care keep both pieces flat. Watch the pointy end of the shears and not back where it is cutting helps stay on the line. You can make the job easier by cutting large pieces on the living room carpet since it does not scratch the metal, is easy on the knees and always cooler/warmer than the garage. Make sure your spouse is not at home when you do this.

Bending Wing Skins

The leading edges of wings are rounded but the radius is usually a bit too small to just place the skin on and bend it around the nose ribs. Therefore, the leading edge of the skin must be pre-formed to make a better fit. This version of the leading edge forming involves some 1.5 inch PVC pipe (or other suitable diameter) and a vacuum cleaner. The PVC pipe is drilled in several places to allow the air to be evacuated. The sheet is sealed with duct tape and plastic sheets with the PVC pipe inserted full length. One end of the pipe is plugged, a vacuum cleaner hose is connected to the other end. Here are photos of the set-up before (left) and after (right) adding vacuum. Once everything is sealed, the pipe is double checked to be centered and straight and the vacuum is applied to the PVC pipe opening. Within about 30 seconds the skin will looks like in the right picture.

The skin looks quite flat and the leading edge will pull slightly away from the PVC pipe. After removing the vacuum and the wrapping, the skins will have a nice rounded bend and the short portion will stand up at about 80º to 90º. One advantage over bending the skin in a brake is this method will not leave any marks on the skin.

Jewelers Saw

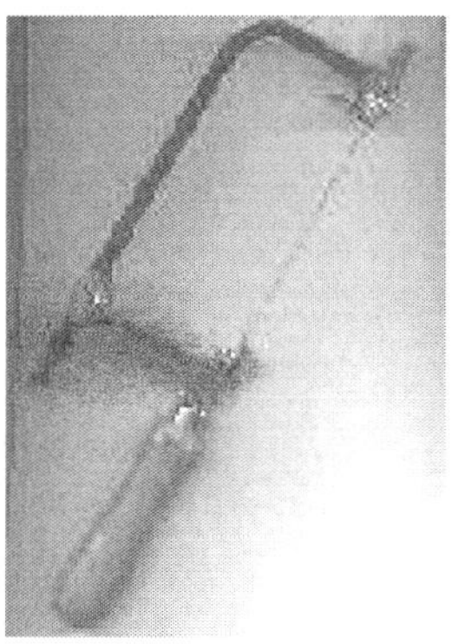

If you need to cut small parts with intricate shapes from metal, a band saw may not be the ideal tool. Bandsaws are best for straight cuts or large radius and unless your bandsaw speed can be adjusted to slow speed for cutting steel, you may need a different method for cutting steel. A jeweler's saw is a very accurate, but slow alternative method. These come in various frame depths, with a 5 inch depth fitting on most occasions. The blades are about 5 inch long but are available in different lengths. The size describes the thickness of the blade and the number of teeth. A #2 blade works about right for a thickness from 0.032 inch to 0.090 inch of steel. A higher number, such as #4, works fine for 0.125 inch. Install the blade with the teeth pointing down and make sure it is nice and tight. This works best for most people. When sawing, it is important to keep the saw straight and not to angle it. This will assist in keeping the filing of the metal down to a minimum. Use long, even strokes instead of short, quick ones. It will cut quicker that way. Be generous when lubricating your blade, bee's wax works well. Use an ultra-fine point sharpie to mark the line on the metal - and the blades are thin enough to cut right on the line. Clamp the work so it extends over the edge of the table, and use a vertical motion of the saw frame. Just remember to only pull straight down, this saves a lot of broken blades. You can cut a 90-degree angle if you keep the blade moving a bit while you make the turn - the blades are about the same thickness as the kerf. A few passes with a file finishes the piece.

Avoiding Scratches

To reduce the chance of marring or scratching a piece of sheet metal when drilling holes, apply a layer of tape over the area of the holes. Drill the holes and remove the tape prior to riveting. Tape can also be used on clamps holding the sheet metal.

Another method is to use a small pad on the drill itself. This can be rubber or leather. This prevents the drill from scratching a circle around the hole if it contacts the part, especially if it is aluminum. When using this, check that no debris or chips adhere to the rubber pad which could also scratch your work.

Form Blocks Materials

There are a lot of opinions on the type of wood to make the formblock from. Everyone agrees that soft woods like pine or fir are not acceptable because they would not stand up to the repeated hammering necessary to form the multiple ribs. Several have made their formblocks out of a material called MDF manufactured wood. MDF stands for Medium Density Fiberboard and comes in 4 ft x 8 ft sheets like plywood. Many builders have used hardwoods like Oak, Mahogany, Ash or Birch. These may be hard to find in the thickness needed and are expensive. These woods can change size and/or shape with humidity changes. MDF will not do this and when drilling it you do not have to worry about the drill bit following the grain and getting off center. A formblock should be at least 3/4 inch thick, 1 inch is even better. Two pieces of the MDF glued and clamped to form one piece with 1.5 inch thickness should be sufficient for all applications.

Forming Ribs

Rib blanks are cut from aluminum sheets and the rib flanges are then formed by bending them over the edge of a formblock using a dead blow hammer or rubber mallet. The finished flanges of the ribs should end up at a 90 angle to the rib web. One of the attributes of bending structural aluminum is that it tends to "spring back". Meaning that if you form the edges of the jig at a 90º angle, then bend the aluminum over it, no matter how much you hammer it, the aluminum will spring back a small amount and the bend will result in a more than 90 degree angle (about 95 to 100). The spring back seen with 2024 aluminum is the most pronounced. It is almost non-existent with 5052, which is a lot softer than the 2024 aluminum used for the ribs.

To account for "spring back" while forming the ribs, use a disc sander table, set it up to have a 6 angle from the disc and sand the edges of the formblocks to have an 84 angle edge instead of 90 . This will result in rib flanges closer to the required 90 angle (after spring back of the aluminum) when the flanges are formed. Even if your ribs are all the same size, you will need two formblocks, each with the edge angled at 84 but in opposite directions, one left and one right. You have to make left wing ribs and right wing ribs. This means that half of the ribs have the flange facing left and half of the ribs have the flange facing right (left wing, right wing).

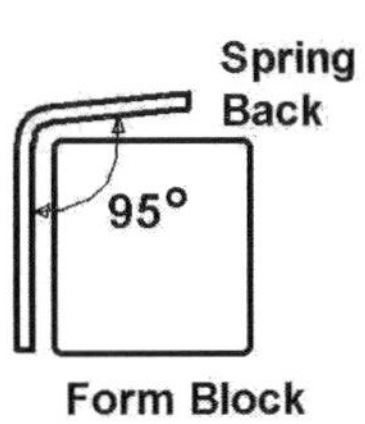

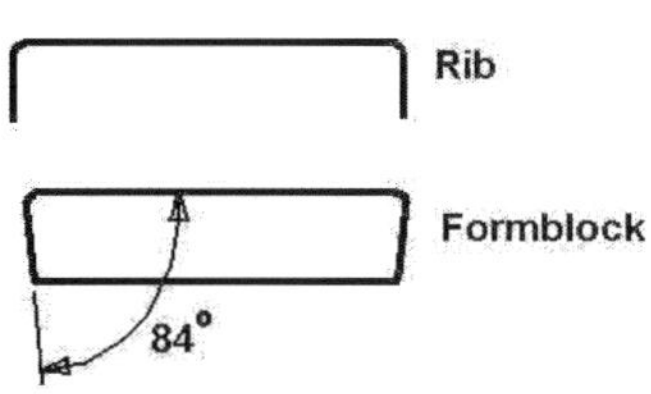

To bend the flanges, the rib is clamped between the two formblocks, bolted together through the jig holes to assure proper alignment and clamped down tight. In the above picture is a nose rib clamped in the formblocks just prior to bending the flange: The marks on the formblocks are where the rivet locations will be. These marks will be transferred to the nose rib after bending the flange. That will give you a reference of where not to apply the flutes for straightening.

The most difficult part of the rib flange is at the leading edge because of the small radius of the airfoil shape. This is where damage is most likely to happen. It helps to reduce the width of the flange at the leading edge or eliminate it altogether. Go slow and be careful when you work on this area.

Cutting Large Holes in Small Parts

If you need to cut a large hole in a small part, do this before cutting the small part from the big sheet. This picture shows an example. The builder had to cut several identical parts from a larger sheet. It is a lot easier to cut the holes first while there is a larger piece to clamp or to hold on to.

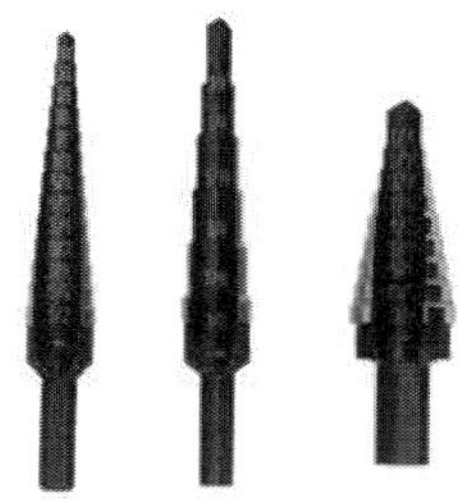

To drill a large hole (>3/16 inch), it is best to pre-drill the hole with a smaller size drill first. Most larger drill bits do not cut well unless they are used to enlarge an existing hole of about 1/3 of their own diameter. This is due to the core of the drill bit, which limits the length of the most effective cutting edges unless a special cut is used. It is often quicker to use a step drill, which progressively opens up a hole until you get to the right size. Just make sure you do not drill too far. It may help to mark the step which should not go through the hole with a piece of tape or a marker, for example.

Bending U-Shaped Brackets

With a normal brake, you cannot bend narrow U-shapes bends. To fabricate such brackets, use a bending block. It can be made from wood or plastic but metal is best. It should have the inside dimensions of your bracket. Break the corners; they will determine the bend radius. The pictures show the bending block and the flat pattern of the part. The bending block is clamped in a vise with the part and the flanges are hammered until they are folded over. Steel springs back very little, 90 bends can be made with a square bending block.

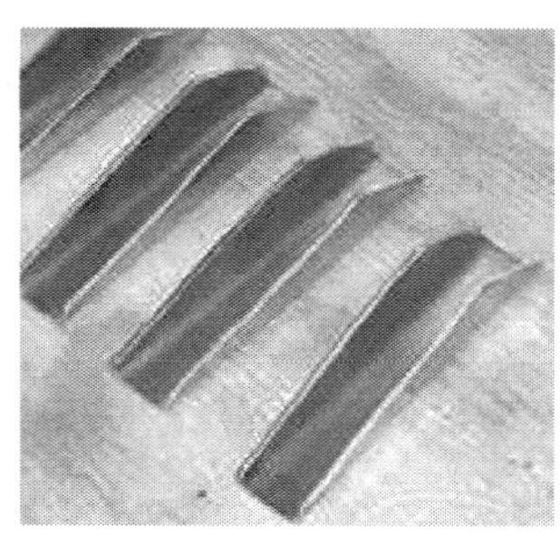

Bending Trailing Edges

You can make a bending jig using two pieces of wood and some hinges. To avoid bending the trailing edge into too tight radius, put a small wood dowel into the bend of the skin while you work the bend. To get enough leverage to bend the trailing edge, clamp two bar clamps onto the top upper piece of wood to give you more leverage and clamp or screw the bottom piece of wood to your workbench. If the part is long, you may need to use extra C-clamps in the middle to get even pressure across the full length of the trailing edge. Bend slowly, and check often until you get a good fit. The aluminum skins should lay flat and just touch the spar without you having to push. Do not over-bend, or your skin will want to bow in instead of laying flat.

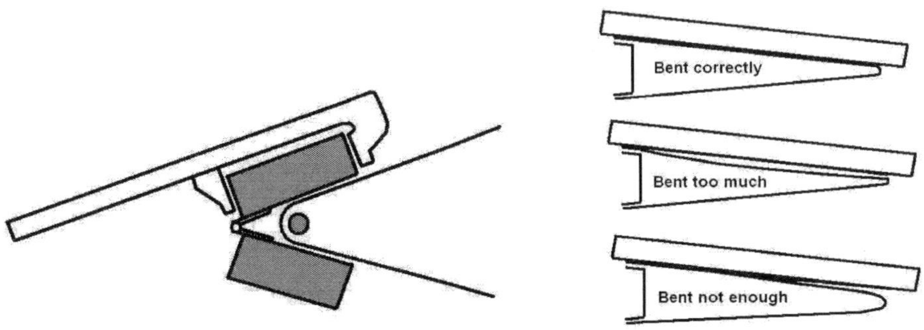

If the bend in correctly, the straight edge lays flat along the entire length of the skin and the bending radius is correct. If it is not bent enough, the edges will not match up to the web and the bending radius is too large. If it is bent too much, the radius may be correct but the angle is not and the edges are too tight. The part will be bowed in.

Deburring Sharp Edges

This is a very important step. Do not be in such a hurry that you skip it. By dressing all the edges of the aluminum parts, you remove potential stress points and ensure proper fit. Just about any cutting process leaves little burrs or sharp edges on metal pieces, and leaves little tabs on the thicker parts. All of these imperfections need to be removed. A good rule of thumb is to run your fingernail along the edge of a part. If you can feel the bumps, you need to do more edge dressing. You should be able to run your finger over each edge and not cut your skin. Include the inside of lightening holes, they need de-burring too. Tools used for deburring include burnishing tools, or edge de-burring tools. Many people prefer a file if there are heavy burrs, and finish up with emery cloth. As an alternative, smaller parts are easy to edge dress using a Scotchbrite polishing wheel mounted in a bench grinder. A few other useful tools are a flapper wheel mounted in a drill press or a vixen file cut down and mounted with a wood handle. If you use hand shears to cut the metal skin, there is no good way to get rid of the ripples.

Dimpling

Dimpling is required for flush-riveted aluminum sheets which are too thin for countersinking. A dimpling tool set consists of a male and a female part. You can easily make the female part yourself. Take a thick steel sheet, drill a hole, countersink it and it is done. In the setup shown here a long reach dimple tool is flanked by two carpet covered blocks to protect the skins on either side of the tool. The hammer is a 1 lb dead blow mallet. The long reach dimple tool should be able to reach most of the rivet holes in the skins.

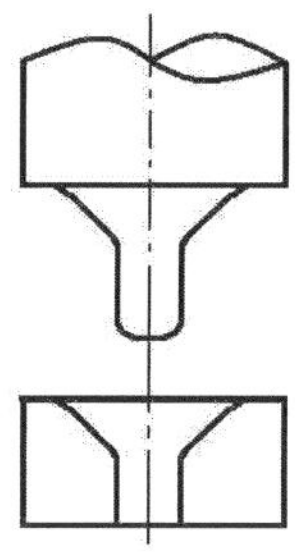

To use the long reach dimple tool, the shaft needs to be raised, the skin is positioned and the hammer is used to hit the shaft. If you have large skins, it would be helpful to have a third hand to accomplish these tasks. A neat modification to the C-frame makes dimpling skins a lot easier. It allows you to use both hands to reposition the skin and gives you a handy place to put your hammer. A bracket is added which pivots to raise the shaft when you place the hammer on it. When you pick the hammer up, it lowers the shaft into position ready to be smacked. The bracket can be made from thick aluminum sheet stock or whatever material you have around. Cut and bend it, drill the hole for the pivot (1/4 inch) in the C-frame and bolt it on. The left picture shows the bracket and the shaft raised with the hammer to act as the weight.

Tip for dimpled holes: *very slightly* countersink dimpled holes - by hand (only two or three "half turns") to get a real nice, flush seat of the flush rivet. "Spring back" dimple dies will reduce or eliminate the need to do this. As with most of these techniques, experiment to find the best results.

Clecos

A cleco is a tool which is used to hold sheets with holes together prior to riveting. A cleco hole is by design undersized and after every thing fits right it is redrilled to proper rivet size and riveted. There are several types of clecos for different situations. Most are put into holes to keep things lined up while you are busy drilling, fitting and riveting. Along with clecos, you need cleco pliers. One jaw of the pliers has a U-shape to fit around the barrel, the other a flat end to depress the plunger. To insert a cleco, use cleco pliers to depress the spring-loaded plunger. This relaxes the cleco pin and reduces its diameter allowing it to fit into the hole. Once the pressure on the plunger is released, the pin expands and fills the hole tightly. On this fuselage the skins are attached with clecos, making it look more like a porcupine. The right picture shows a pneumatic cleco tool. If you have to set a lot of them, this can be useful to have to reduce the strain on your hand of repeatedly squeezing the pliers. Otherwise manual cleco pliers will do the job.

Clecos are color coded, with copper for 1/8 inch diameter holes and brass for 3/16 inch holes, for example. There is a different style of clecos for clipping the edges of two pieces together. Clecos are generally **not** put into every hole - just enough to hold things together in a proper manner. About every third or fourth hole is plenty, otherwise you would quickly run out of clecos. When you are ready to rivet, start in the center holes, and work your way outward. This helps to keep things lined up a bit better. Remove the clecos as you approach them for either riveting or drilling - if drilling, keep moving the clecos to a hole you have recently drilled to size.

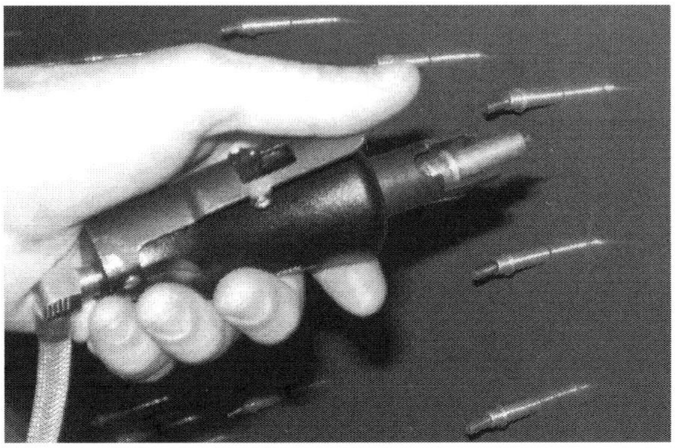

Countersinking

Drilling a hole and then countersinking it can result in a countersink that is not concentric with the hole. If it is significantly off center, it can unduly stress the hole and the screw. To avoid this problem, you can use a combination drill bit and countersink cutter. This picture shows an adjustable countersink. The countersink cutter can be moved to allow different hole depths before countersinking occurs.

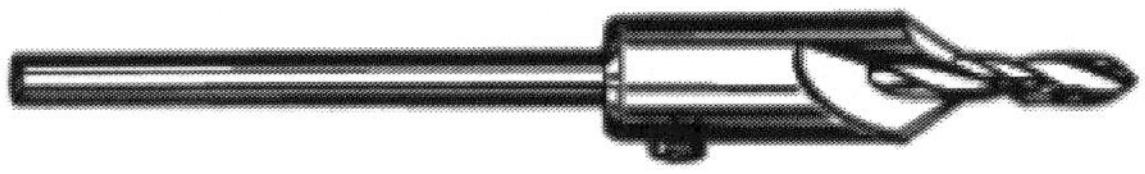

A cross-hole countersink (left picture below) is a cone-shaped tool with a cutting edge provided by a hole that goes through the side of the cone (see picture at right). The intersection of the hole and cone form the cutting edge on the tool. These tools are best used as deburring tools, where the burr from a previous machining operation needs to be removed. They may be used in softer materials (such as wood or plastic) to create a countersunk hole for a screw.

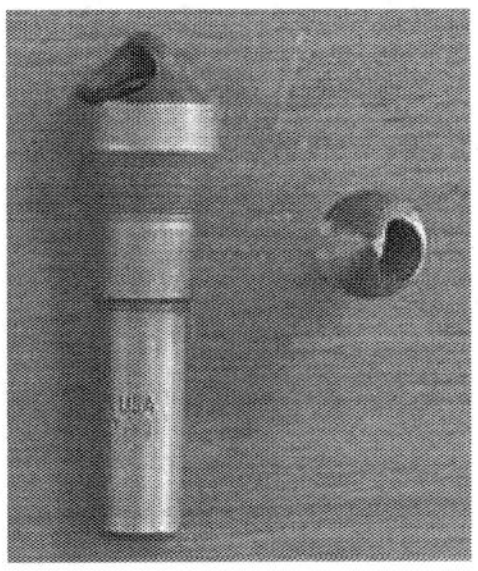

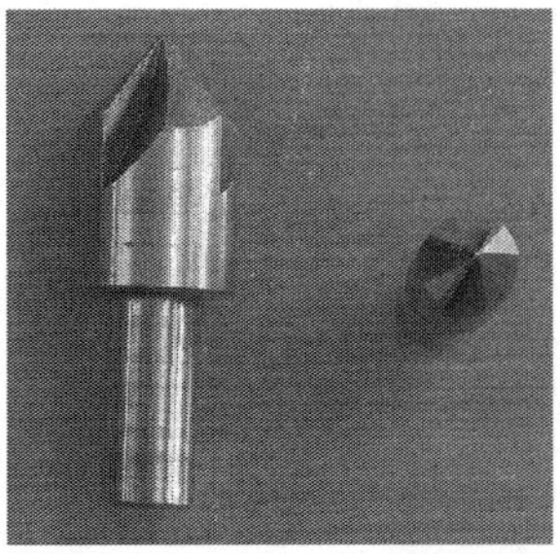

A fluted countersink cutter (center picture) is used to provide a heavy chamfer in a drilled hole to allow the correct seating for countersunk head rivets or screws. The right picture shows a countersink with a stop, which lets you control precisely and repeatedly how deep the countersink will be. You can use a scrap piece of thicker aluminum to help set up the stop. Put a bit in the stop, and the stop into the drill. Then set the stop by eye to what looks right, or just a little under the correct depth, and drill a test countersink. If it was too shallow when tested with either a rivet or a dimpled piece of the correct thickness of aluminum, reset it a little deeper and repeat until it is right. On subsequent setups, you can start setting the depth by pushing the bit into a correctly countersunk hole and set the microstop to be flush with the surface. Then test it on a piece of scrap and make adjustments if needed.

Countersink cutters are manufactured with six common angles, which are 60°, 82°, 90°, 100°, 110°, or 120°, with the two most common of those being 82° and 90°. Countersunk-head screws that follow the Unified Thread Standard often have an 82° angle, and screws that follow the ISO standard often have a 90° angle. Aviation hardware typically has an angle of 100°.

It can be difficult to avoid chatter when cutting with countersink cutters. The shorter and more rigid the setup, the better. To reduce the chance of leaving an undulated surface, good quality fluted countersink cutters sometimes have the flutes (or at least one flute) at an irregular pitching. Too light a feed tends to increase chatter risk. An appropriate response to the chatter may be to decrease speed and increase feed.

Drill and Hole Tolerances

Standard size drills and reamers are not necessarily the size they claim to be, i.e. a 1/4 inch drill bit will make a .250 inch hole +.003 -.001 assuming the bit is new, and sharp, and most importantly drilled in the proper manner. If it is worn, this tolerance can change quite a bit. If you are drilling an important hole where the size matters, such as on a spar, you may want to drill a test hole in a piece of scrap material with the drill bit you want to use and measure the hole to see what you get. Standard bolts, including AN-hardware may be slightly undersized. This undersizing would be detrimental on a stressed fitting such as a spar attachment. No matter how much you tighten the nut, it will move in a shear motion. The result is fretting and eventually a failure of the joint. A close tolerance bolt is no stronger than a standard bolt. It's strength lies in the "close" fitting of bolt in relation to the hole.

Drilling Rivet Holes

When drilling rivet holes, never drill in separate pieces of metal by measuring, always "match drill", meaning drill through all the pieces of metal at the same time. This way it is assured that the pieces will fit together. You can pre-drill rivet holes undersized in one part, then fit the parts together and drill them to the final size together. Wide masking tape or duct tape is useful to initially align pieces prior to drilling. After you match drilled the first hole, insert a cleco to hold the parts aligned. Be prepared to use lots of clecos to keep the pieces of metal matched and flat.

Drilling "from inside out" such as skin to rib rivet holes will help keep rivet holes in the center of the rib flange. Limited access may make this difficult, but if you mark and drill the end holes this way, then after confirming that the rib is straight, you can drill the remaining holes "from the outside in" (normally). When drilling "from the inside out", have a helper hold a block of wood against the skin to hold it against the rib while drilling. This keeps the skin from deforming when you apply pressure and makes drilling easier. This is a good place for those 6 inch and 12 inch long drill bits.

Compromised Edge Distance

The plans for your project may be great, but beware if they say the size to cut the pieces is the final size, which too often leaves no room for deviation. After the first part you make that does not have the required edge distance, you will realize that it is better to leave the pieces a little larger than the plans call for until all the holes are drilled, THEN cut it down to final size.

The rule of thumb is that the distance from a fastener hole to the edge of the part must not be less than two hole diameters. If it does not meet this minimum, make another piece. If it is a large piece that you are trying to save, it may be possible to double up on rivets with proper edge distance or install a doubler piece. Use a scrap piece of aluminum that is at least as thick as the piece with the compromised edge distance.

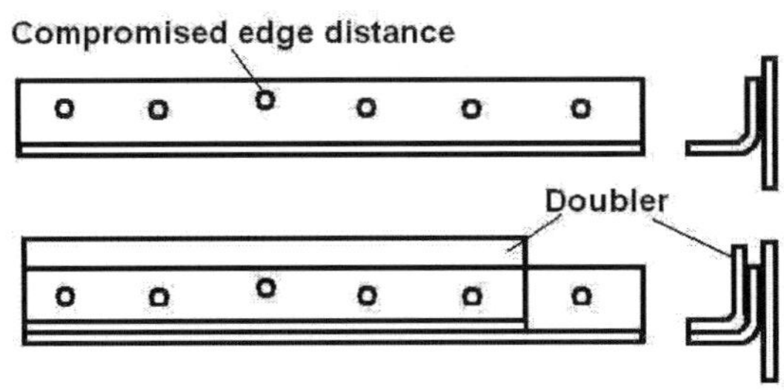

Make the doubler extend to at least two rivet holes on either side of the questionable hole. Then when you rivet it all together, include the doubler on those five holes and you are good. If you have any doubts about the repair, the best solution is to contact the designer or kit supplier for advice.

Riveting Tips

The most important thing is to keep the set and gun parallel to the surface on which you are working. If it is off, you will get a smiley face on the skin, or ruin the rivet and have to drill it out. If you can't easily reach both sides, get someone to help.

Use a hand squeezer whenever you can, it can save a lot of time. By using a pneumatic squeezer and the adjustable set holder, you can finish the work a lot quicker than by bucking. The shape of the parts will determine from which side you will squeeze. Just make sure you have pressure on the factory head before the squeezing pressure starts. Otherwise the rivet may back out, get squished and will be ruined. This picture shows a selection of hand rivet squeezer tools.

Occasionally, in the middle of a riveting session, you will notice that the two pieces of metal you are trying to put together are not tightly together, that is, a space exists between the parts. The quick way to eliminate this space is with the following steps.

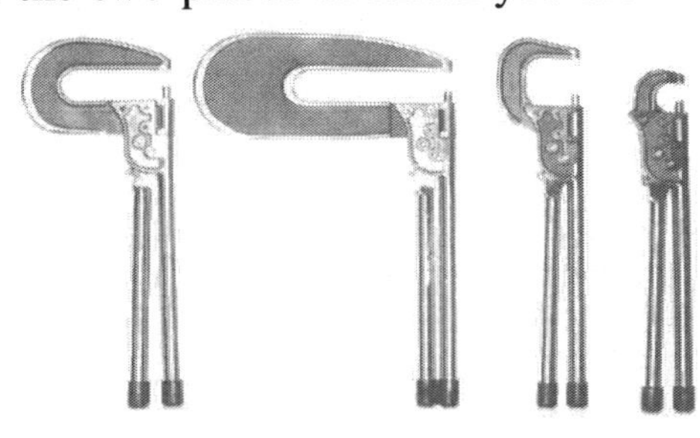

1. Drive and buck the rivet with a couple of taps only.
2. Place the bucking bar right next to but not on the rivet.
3. With the rivet gun on the rivet, tap the rivet a couple of times.

This should draw the parts together and they can now be riveted as normal. If this does not work, repeat the steps again.

A better method, which is a little more involved, requires a special tool. Instead of using the normal bucking bar, make a slender bucking bar with a hole slightly larger than the rivet shank in the end. This way there is support all around the rivet when seating it in the hole. This bar should have a small radius on the end so that no harm is done if it is not quite straight. A countersunk version of this bar allows the same to be done with flush rivets. It is not recommend using the rivet gun to tap them down, because there is a good chance to "smile" the skin that way. Use a plastic hammer to tap down the rivet head. You can also put a small rubber o-ring on the tail of the rivet. It should force the skins together but still squish down to allow a correct shop head to form.

Back Riveting

One thing you may notice on some flush-riveted metal airplanes is the way the sheet metal of the fuselage has been dented by the force of the rivet gun, if used from the outside. The way to reduce or avoid this is called "back-riveting". This means using the bucking bar on the outside of the structure and the rivet gun on the inside. It may not be possible to back-rivet everything from the inside due to lack of space, but do as much as you can from the inside. If you use rivet tape or masking tape over the tops of the flush rivets you can eliminate bucking bar or rivet set marks on the skins and rivet heads. Back riveting takes a little practice, but the results are well worth the effort. You will get the smoothest finished heads by using a backriveting plate. This is a heavy piece of metal that you hold flat against the part with the countersunk head. Hit the shop head with a special backriveting set (left picture).

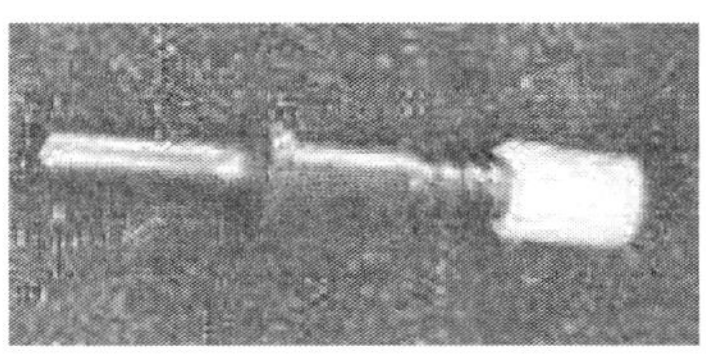

If you cannot use the plate, the next best result comes from using a backriveting bar (right picture). This is a heavy, wide bucking bar that you place against the flat, factory head. Again you drive the shop head with the same special set. This works especially well for the top skins of the wings. Do the top skins first, then buck the bottom skins. This way, the prettier side is up. You can also use this method for much of the fuselage skins.

Aligning Ribs

This tip is for assuring proper alignment of the ribs to the spar. I think the best method to get all the ribs all lined up properly is the string method. Run a string through all of the rib jig holes, pull it tight and make sure the string is center in the jig hole before clamping and drilling. Do **not** use a laser for aligning parts through small holes. The laser light will be bent (diffraction) and it is impossible to know how straight the parts are.

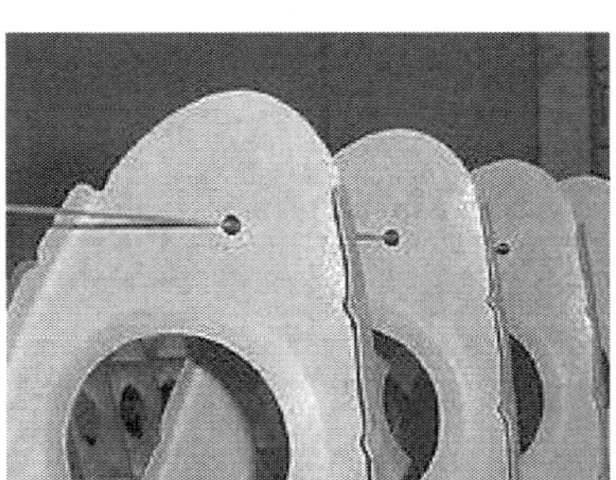

To align the ribs on the spar, use a couple of scrap pieces of aluminum of the same thickness of the skin and some side grip clamps. Simulate the wing skin by clamping the pieces to the spar flange and the rib as shown below: This aligns the rib perfectly to assure a nice smooth transition from rib to spar when you later skin the wing.

Skinning a Wing

Place the skin on the leading edge of the wing skeleton and strap it in place.

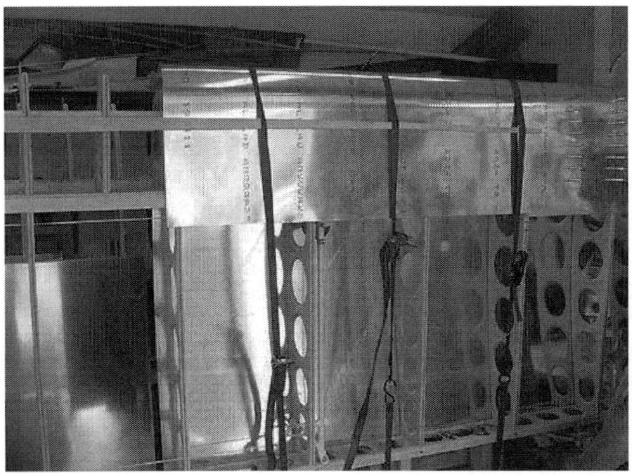

Ratcheting tie down straps work really well for this. A piece of scrap wood was used to help push the skins against the ribs a little better on the bottom side of the wing. Also, a couple of 4 x 4 were used as spacers between the rear spar and the tie-down straps to prevent bending the wing skin at the bottom. Once you are satisfied that the skin is properly positioned and square, mark the rib locations on the inside. Use a Sharpie felt marker, you can reach inside through the lightening holes and trace the actual ribs and spars onto the skin as shown here in the right picture. Once the skeleton is traced to the skin, it can be removed and placed on the worktable. Using a carpenter's square, the nose rib rivet lines can be drawn on the skin at exactly 90 degrees to the spar and centered on each rib. The rivet spacing should be according to your plans and the holes can be pre-drilled in the skin. With skin back in place, the ribs could be nudged one way or the other until the previously drawn lines can be seen in the pre-drilled skin rivet holes. Starting at the top and working down, drill and cleco the skin to the ribs. To pull it down evenly, drill the top row of holes all the way across, then the next row, the next row and so on until reaching the spar.

Skin Lap Joints

The metal of many aluminum skins is so thin, that when you rivet them together, the edge of the outside skin will curl up. To prevent this, put a very small bend about 1/4 inch from the edge of the skin. If you have a break, you can use it for the straight edges, but it is easier to use the vice grip type edge-rolling tool. There are two edge roller tools available. One of them is round and the other one looks like a small pair of vice grips. The two rollers are shaped to give you the proper bend at the proper space from the edge of the skin. They are adjustable to the thickness of the skin being worked. Although you can use a break for the few straight skin edges, this tool is a lot cheaper and works for all the edges. It will take a few test pulls to figure out its use. If you set the tool too tight, it will not roll although it does not necessarily have to roll, as the shape of the rollers does all the work. The only other thing to pay attention to is that the tool does not roll off the edge as you work your way along. There is a lip on the rollers. Just keep that lip up against the edge of the skin and it works well.

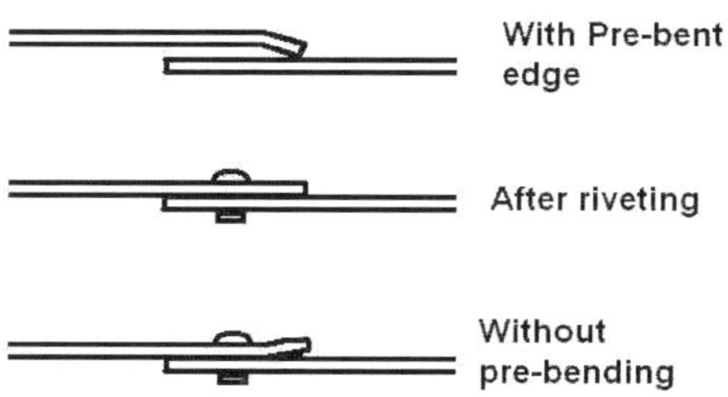

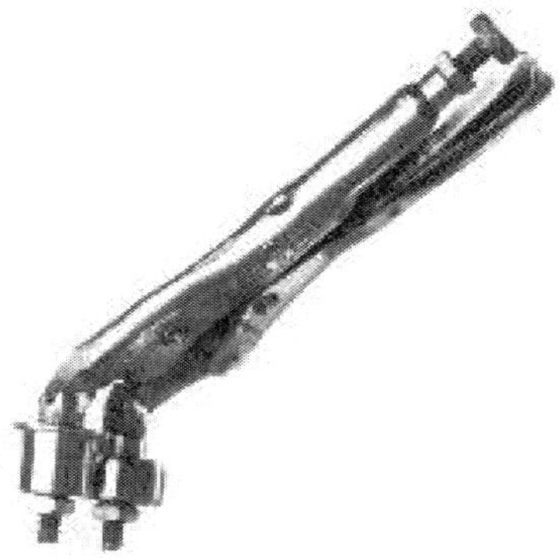

Make sure you do this step before you dimple. The dimples will get in the way of bending, but the bend will not interfere with dimpling. You should do this to the outside edge of every skin unless it will be covered.

Nutplate Installation

Nutplates are important for fasteners in any place where access will later be limited. If you have a lot of nutplates on the plane, it is useful to have a fast and accurate way to install them. I have seen several methods for installing nutplates. Here is one of them:

1. Drill a hole where the bolt or screw will be (may be undersized). Using the correct size cleco, attach a nutplate to the hole using the cleco through the center threaded portion (where the screw or bolt will go) as a template. If you do not have clecos, you can also use a short screw or bolt to hold the nutplate in place. It just takes more time to use a screw instead of a cleco.
2. Use a long drill bit of the size appropriate for the holes in the legs of the nutplate, drill one of the legs and put a cleco in that hole.
3. Drill the other leg with the long bit.
4. Remove the clecos and the nutplate. Now drill the center hole to the proper size for the screw or bolt you are installing.
5. Deburr, prep and prime the parts.
6. Dimple or countersink (depending on the thickness of the material) the holes where the nutplate legs are.
7. Attach the nutplate with flush rivets.

If you drill a lot of nutplate holes in a row, do not use the same nutplate as a template to drill the holes with more than a few times. The holes in the nutplate will get enlarged after being used over and over and the accuracy of the holes will suffer.

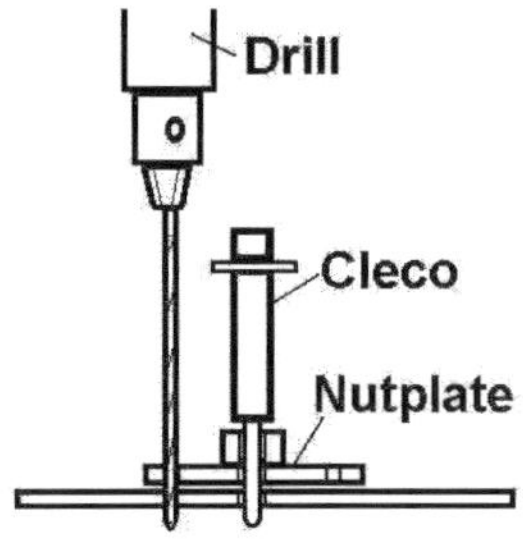

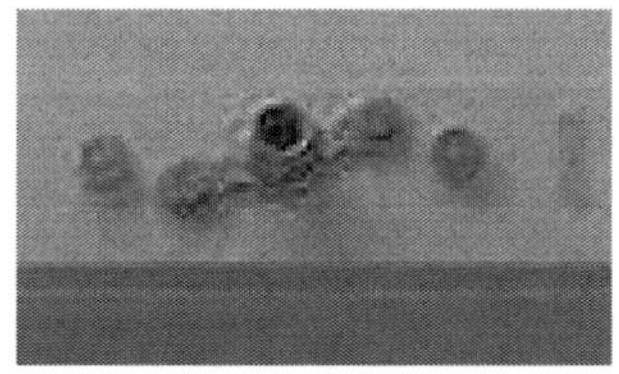

Flanging the Lightening Holes

The plans may call for metal ribs to have lightening holes. To stiffen the rib web and reinforce the hole edges, the holes should have flanges around each hole (for example 3/8 inch wide flange at 30). There are several ways to accomplish this. One method is by using the two formblocks and a die block made from the MDF wood. For the ribs, use the two form blocks which were used to form the flanges. Cut two (or the number of your lightening holes) large holes in each of the two formblocks that are 3/4 inch larger than the lightening holes. To make the flange, start with the double thick piece of MDF. Using a flycutter, cut a plug for each hole out of the double thick MDF that exactly fits the holes just made in the formblocks. Turn this plug on a lathe to produce a 40 angle at its bottom edge, where it will make contact with the exposed rib material in the hole. The 40 angle allows for spring-back of the aluminum and gives a nice 30 bend. Clamp the rib between the two formblocks and bolt them together through the jig holes. Clamp it also in the center between the holes to assure a nice tight squeeze.

The pieces are then placed into the bench top press and the plug is pressed down against the rib (bevel side down). When done, you have a nice 30 flange that is 3/8 inch wide around the lightening holes in the rib. The only problem is that there is now a bow in the rib lengthwise as can be seen in the picture above. To remove the bow, the pressure around each hole created by the bend of the flange, needs to be relieved. To do this, take some scrap metal bar (1/4 inch thick x 1 inch wide) and smooth and round the end. Use the bar with a small hammer to go around each hole right at the bend line and tap - tap - tap while pressing down on the rib to flatten it out. It takes a little trial and error and some practice to get it just right, but when done the rib lays nice and flat again.

Drill Template

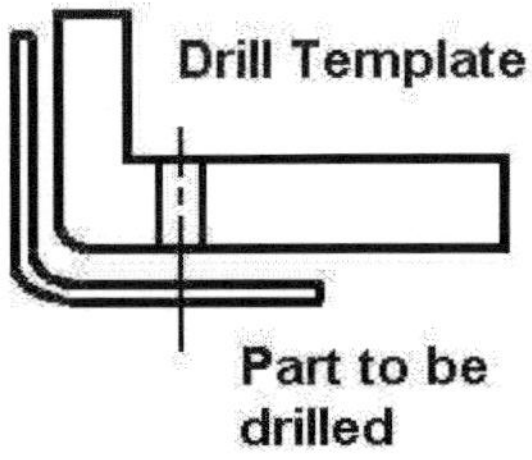

If you have a lot of pieces which all need the same hole pattern, such as angles, it may save a lot of time to make a drill template instead of marking and drilling each piece separately. In this example a piece of the angle is measured and drilled. The holes need to be measured in such a way that when another stiffener is placed underneath, the holes are centered on the stiffener and properly spaced. They are then clamped together for drilling.

Radius Template

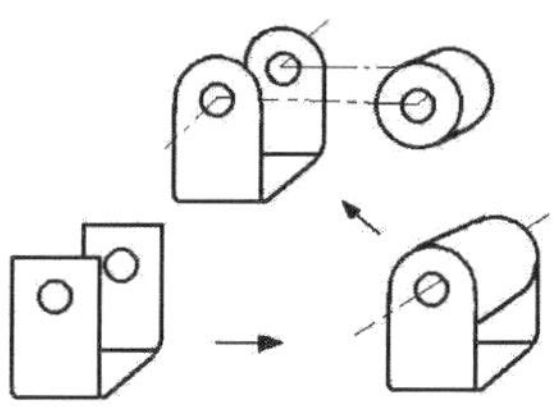

It is easy to use a round spacer to help you file a perfect radius on your fitting. Select or make a steel spacer with the radius that you want your part to have. The diameter of the inner hole should match the size bolt you will use. Clamp the spacer between the ends of the U-bracket you want to shape by inserting a bolt in the hole. Use a file to remove the material until you are down to the profile of the spacer. The resulting radius will be concentric with the bolt and the same on both sides.

Cutting Tubes

You can cut tubes with a saw, with a cutting blade in a die grinder or with a tube cutter. The tube cutter is especially useful for small tubing or soft material where you want a straight cut. It does not produce chips which could contaminate the inside of the tubing and does not require clamping which might deform the tube. The tube cutter is positioned on the tube where you want to cut it and is adjusted to the diameter of the tube. Rotate it around the tube while the sharp blade cuts deeper and deeper into the tube material. Continue to tighten the adjustment screw until the tube is cut. The ends of the tube will be perfectly square.

If the tube is too large for a tube cutter and you want to maintain a perfectly square end, use the following trick. It's easy to do if you wrap a piece of paper around the end of the spar and grind/sand right up to the paper.

Tube cutters also have drawbacks. The tube cutter can roll the end of a soft material tube such as aluminum inward to the point that it takes a lot of tube material removal, not just deburring, to get to a full tube inner diameter end. Compressing the tube material inward during the cutting process with a tube cutter can work harden the material and this can lead to subsequent cracking. Remember that if the end of the tube is going to be flared the material at the end of the tube will be worked again in the opposite, outward, direction.

The alternative method is to cut the tubing with a fine tooth hacksaw or band saw, square the tubing end up with either a sanding disc or file, deburr, then flare. A tube cutter is still a handy tool to have. You can cut the end of the tube a bit long, then square the end up with a sanding disc or file to virgin material, deburr, and then flare.

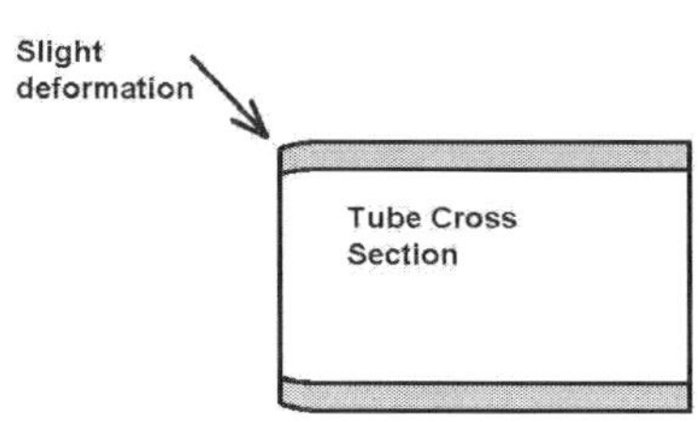

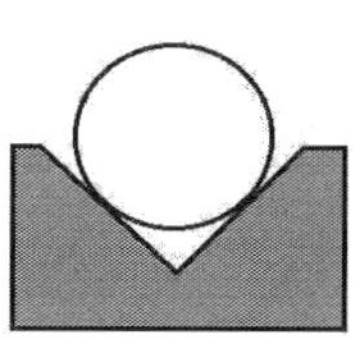

Support for Tube

Clamping or supporting a round tube securely for a marking, cutting or drilling operation without deforming it can be tricky. A very simple support looks like this. You can buy it or can quickly make it from a block of wood. It keeps the tube from rolling or turning.

Tube Bending

The radius that a tube can be bent without kinking or otherwise deforming it depends on the tube material, diameter and wall thickness. The softer the tube and the smaller the diameter, the easier it is to bend it tightly. For all but the largest radius you should use a tube bender. This can be a commercially available device or a simple wooden form with the correct radius. A tube bender which supports the sides of the tube to reduce ovality can only be used for one tube size. Large diameter tubes with thin walls like exhaust tubes can only be bent with mandrels. In that case it is easier to buy pre-bent pieces and weld them together. Poor man's mandrel bending goes as follows: Cap one end of the tube and completely fill it with dry sand. Cap the other end and bend it. The sand minimizes the cross section collapse and ovality. To further reduce the bend radius, heat the tube to slightly soften the metal.

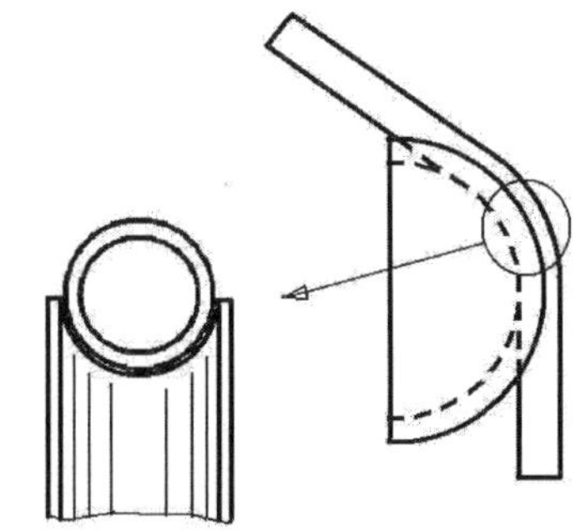

Tube Notching

There is an inexpensive, accurate tube notching helper which you can find on these web sites: http://www.metalgeek.com/static/cope.pcgi and http://metalgeek.com/static/dan/. You can download a small program (TUBEFIT) which calculates an accurate profile of the cut for your tubes. This profile can be printed to scale and can be used as a template for cutting the tube. Just mark the tube from the template with a felt tip marker, and make a saw cut in from each side. The cuts are almost straight lines, but not quite. Cut far enough that the cuts intersect. This will leave two sharp ends that can be trimmed off. The whole ID of the cut tube will then fit against the uncut tube, and there will be a sort of bevel between the ID and OD which is just right for welding.

If you need to cut tubes on a regular basis, consider buying a tube notcher. They generally do well with larger tubing (1 inch diameter and up). But for a few tubes (= one airplane), there are a few methods which will get you by without spending a lot of money. For material under 0.050 inch thickness you can use aviation snips. For thicker material, you have choices of a vertical mill to files to die grinders. The die grinder can be fit with various sizes of 2 inch long drum sanders. Common diameters are 3/4 inch and 1 inch drums. Get a selection of half round and round files from 12 inch to 6 inch. Have your grinder and belt sander close to your jig; you can rack up a lot of miles walking back and forth to dress a tube.

Chamfer a Bolt

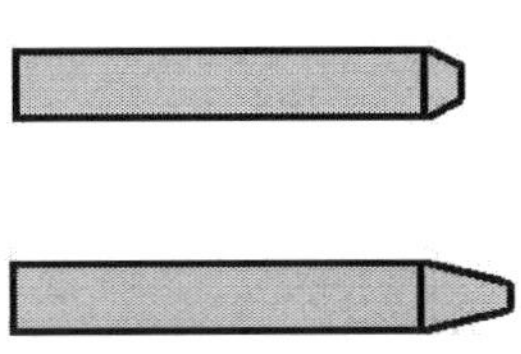

To make a "bullet head" on a bolt or steel shank, put the bolt into a drill, run the drill in the opposite direction of a belt sander and hold the bolt at the desired angle against the sander. You can convert your drill press into a "lathe" if you spin the bolt in the drill press and use a file or other tool on the bolt to shape it. With this method and a saw blade, you could for example cut a grove into it.

Part Marking

There may be a lot of similar looking parts in your kit. What is the best way to store them, mark them and keep them apart? When you clean, prime and paint metal parts, part numbers or markings written on them will most likely disappear. Here are a few possibilities:

- Attach a tag with the part number, or description to the part. This can be a piece of card stock attached with a wire.
- Keep small parts in individual plastic bags marked with the part number.
- Permanently engrave the part number before prime and paint so it will show through the coating. This can be done with any sharp object or an engraver.
- Attach a piece of wire through a hole in the part with a certain code. Wire is attached to either the "up-pointing" or the "forward-pointing" hole in the part. For parts that are reversible, attach the wire to the top-left hole.

Welding

Edges

Before you weld parts close to their edge, consider the effects. If the part is cut to its final dimensions, it may end up short or damaged because its edges melted away. Leave the part oversized for welding and trim it afterwards. Examples are this bracket reinforced with a welded washer or these fittings.

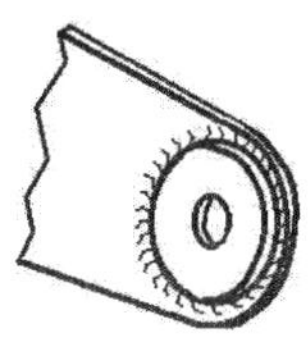

Welding Stainless Steel with a Torch

The properties of stainless steel are such that a few things should be observed when welding this material with a torch. Stainless steel is a poor conductor of heat, relatively speaking. It conducts heat about one-third less easily than ordinary steel. It expands roughly 1.5 times as much as ordinary steel. It also has a slightly lower melting point than regular steel. Since the heat is not conducted away rapidly, stainless steel reaches higher temperatures in the comparatively small area of the weld. Stainless steel has a tendency towards more distortion and residual stresses than ordinary steel. For these reasons, use a smaller flame to cut down on the amount of heat. Too much heat causes warping and buckling.

If the pieces to be welded are thin, that is, up to 1/16 inch (1.6 mm) thick, the usual technique is to turn up the flanges on each edge to a height of about 1/16 inch, paint the flanges with flux top and bottom and melt them down to form the weld. The flanges furnish enough metal to fill the seam and filler rod is not required. If the parts to be welded are between 1/16 inch and 1/8 inch thick, the edges should be joined in a plain butt weld, preferably with a backing of copper under the seam to prevent the molten metal from flowing out the weld. If the parts are thicker than 1/8 inch, the edges should be beveled to provide a V that allows the heat from the torch flame to penetrate the metal all the way through.

The parts should be clamped or jigged so that the line of the weld slants downward slightly in the direction of the welding. This permits the flux, which melts at a lower temperature than the metal, to flow forward and protect the metal until it fuses. One of the precautions taken for welding stainless steel includes avoiding an oxidizing flame. A flame with too much oxygen must be avoided because it oxidizes the molten metal and makes it porous. It is advisable to use a neutral flame in which the inner cone is surrounded by a very small "feather" of acetylene. This feather protects the welder against any variations in gas pressure that might turn a perfectly neutral flame into an oxidizing flame. Use a welding tip or nozzle which is one size smaller than that ordinarily used for a similar thickness of carbon steel. Both the forehand and the backhand methods are suitable for stainless steel, but the latter is recommended for the heavier gage parts. Do not stir the molten metal because that would introduce air and increase oxidation. Weld the pieces entirely from one side; do not shift from side to side. Fill the joint completely with weld metal so that you do not have to go back and patch up a spot you should have finished the first time. You should avoid any tendency to work in towards the center, away from the edge of a joint. If the joint is located so that you cannot avoid working from an edge, you should start the weld an inch or two in from the edge and not at the edge. Weld from that starting point until that particular seam is completed, then return to the starting point and finish the weld out to the edge. As you approach the end of a seam, raise the torch very slightly, thus allowing the molten steel to cool slightly to prevent it from spilling over the edge. The relatively small amount of oxide formed during the welding and any dirt or other impurities that may have been in the original steel work their way to the surface as the welding progresses. When the job is complete, they are in the reinforcing portion of the weld where they can be ground off.

Techniques for Welding Aluminum Alloys

Welding aluminum alloys with a torch is different from welding steel. A neutral flame is desirable for work on aluminum, but as for stainless steel, a very slight excess of acetylene is recommended. In addition to being neutral, the flame should be “soft”. This means that the mixture of gasses comes into the welding tip at slow speed. The flux for aluminum alloy is a powder that is mixed with water to form a thin paste, using two parts of flux to one part water. The flux has a lower melting point than either the aluminum or the oxide formed on the surface of the aluminum. The flux breaks down the oxide film and carries it away, permitting the liquid filler metal from the welding rod to flow onto clean molten base metal. In welding aluminum alloys, hold the torch at a pronounced slant to avoid blowing holes into the metal. For flanged and butt joints, the angle should be about 45 . For lap joints, the flame should be held at about a 60 angle to the surface of the base metal. In welding a T-joint, the flame is held equally between the two pieces. The inner cone should be about 1/8 inch from the metal and it must **never** touch the metal. Both edges of the joint are heated evenly to distribute the heat thoroughly in the weld area. The torch must not be held too long in the same spot or the entire area around the weld may crumble and drop away because aluminum conducts heat very rapidly.

In welding most metals you can tell when the material has reached the melting point by the change of color. This is difficult with aluminum alloys because there is no pronounced change in color when the melting point is approached. It will be solid one instant and then, without any obvious change in appearance, it will melt and flow. With more practice, you may be able to recognize a light-gray color of the metal when it begins to melt if you wear light blue glasses. Also, it is possible to touch the surface of the metal very lightly with the filler rod and thus “feel” the change as it begins to take place. When the aluminum begins to melt, it seems soft and elastic. When this condition is recognized, immediately dip the welding rod into the puddle of molten metal and allow the rod to melt with the metal in the puddle. The filler metal must not flow ahead and onto the comparatively cool and unfused edges of a seam.

Working with Wood

Staples and Nails

A stapler can be used to hold gussets to capstrips while glue sets up. If you use staples to hold pieces of wood together until the glue is cured, it will be much easier to remove them afterwards if you place a strong string or wire underneath them. To aid removal of the staples, pull the wire up.

Nails are also commonly used for this job. To remove them easier after the glue is cured, nail them on with strips of scrap plywood underneath. If there is any risk of this sticking to your parts, add a piece of plastic or wax paper in between. For removal, pull up on the plywood. As an alternative, use small, tapered tack nails. These have the advantage that they do not have to be driven in all the way to hold the parts together. Since part of them sticks out, they are easy to remove afterwards.

Deburring Wood

Even wood can have sharp edges which can cut your hands. You can "deburr" wood after cutting it in a miter box saw with a razor blade in a gasket scraper handle instead of sandpaper. It is quick and easy with no sanding dust.

Bonding Parts

Regardless of what material you want to bond, apply the adhesive to both sides. Apply enough pressure to squeeze some but not all of the glue out. If you cannot squeeze any glue out, there either was not enough glue, the glue is too viscous (possibly too cold) or it is a very large area and not enough pressure. None of these three conditions would result in a properly bonded joint.

Bonding a Facesheet to a Truss

Once you have completely assembled the truss or frame, it will need to be bonded to a facesheet (plywood). You should only apply the glue to the facesheet where necessary. As preparation, place the truss on the facesheet and trace its outline with a pencil to mark up where the glue needs to be applied. Most likely the side of the truss which faced up when you built it looks better. Bond the plywood to the other side so that the nice looking side remains visible. Let it cure with the plywood side down. That way the excess glue will run down and form small fillets.

When installing plywood sheeting on a frame, cut the plywood a little larger. After the glue has cured, use a small block plane to remove the excess plywood around the perimeter. It goes surprisingly fast, is precise and no dust is created.

Varnish

All wood surfaces which are not bonded need to be varnished for protection. The varnish is usually clear, which makes it hard to see where you have already varnished the wood. To make it easier to see, you can add a small amount of colored paint to it to slightly tint it. If you have the access, varnishing the wood after bonding is recommended to avoid contaminating the bonding surfaces. If you have to varnish before bonding, use a pencil mark up exactly where the bond lines will be and exclude these areas from the varnish.

Wood Truss on Table

To keep the glue joints of a wood truss from sticking to the table, put pieces of plastic underneath them. If you have a full scale print of the part you are building, you can protect it with plastic and build the part right on top of it. Most plastic will release readily from the glue. But just in case, you should test it on a small piece. To improve it, you could wax the plastic with a mold release agent. If the plastic sheet is too thin, it might tear when you try to remove it and make it more difficult to get rid of.

Wood Gussets

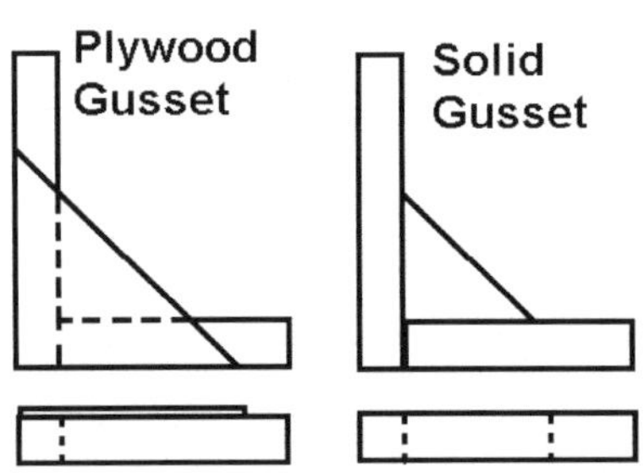

There are basically two ways to build gussets. One method is to glue plywood on top of the truss, the other one is to glue in solid corner pieces to reinforce the corner.

The plywood gusset is lighter and requires less wok to get it to fit well up front. The solid gusset is stronger and has more bonding surface, but has to be cut precisely to fit well. To make it easier to see the grain direction, moisten the end of the block of wood.

Wood Glue

The most common glue for wood is T-88, which is an epoxy resin. It has the advantage of having the simple mixing ratio of 1:1 and is fairly viscous. Most epoxy resins used for building composite airplanes are also suitable for bonding wood. If their viscosity is too low to fill the gaps, you can mix it with a small amount of cotton flox, as a filler. Different resins have different cure requirements. Unless you have a very large surface or a very slow job to deal with, use the fastest curing hardener in order to obtain the best cure at room temperature.

Cutting Thick Blocks of Wood

First of all, have your bandsaw well adjusted. When cutting material from a larger block using a bandsaw, it helps to draw the cut line on the material even if you are relying on the fence to achieve the desired width. If the blade drifts due to misalignment, you will see that you are heading off course before you cut too far.

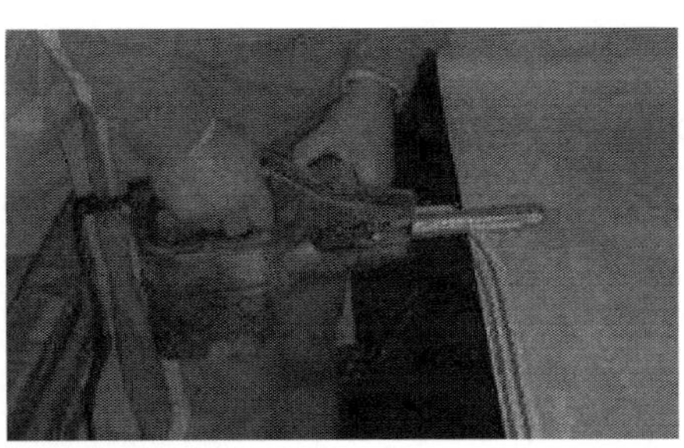

Plywood and Scarfing

The strongest plywood is laminated birch. It is the least expensive but is about 25% heavier than Mahogany faced poplar plywood. If your main concern is weight and not strength or cost, consider using Mahogany. Thin plywood can be cut with a sharp, heavy duty knife. Repeat passes along a steel rule and make long, straight cuts. This is easier than using a saw.

Plywood pieces are joined with scarfed edges. The proper scarf ratio is 1:15 which means the width of the scarf is 15 times the thickness. One quick way to prepare long edges for scarfing is with a handheld belt sander. Get it roughly to the right size with the machine and finish it by hand with a sanding block.

Sanding Wood Ribs

For giving a rib the right height you can make this tool that sands one rib and is supported on 3 other ribs. It consists of a long sanding block with sandpaper glued to one area. To take off high spots on the rib you are working on, use it as shown in the picture. Because the sand paper has a height, apply the same sandpaper reversed in the locations on the other ribs so it will not sand them down as well. Sand with the sanding block parallel to the spar or you will end up modifying the airfoil shape.

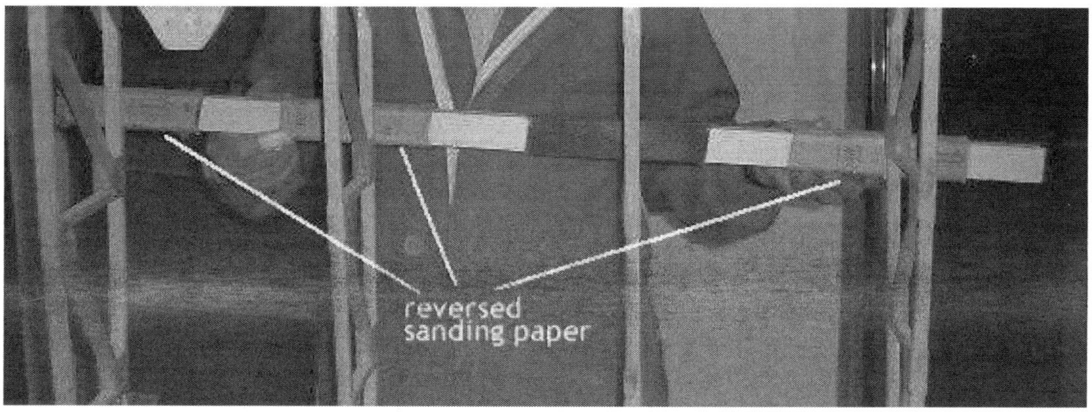

Holding Gussets and Ribs in Place

There are different types of clamps, but this example is a really simple way to hold a gusset in a truss in place until the glue is cured. You can easily make it yourself. Buy small compression springs in a hardware store. Drill a hole the size of the spring into a block of wood. Insert the spring with a nail with a large head as the plunger. A small hole in the back of the block allows you to nail it to the board where you build up the truss.

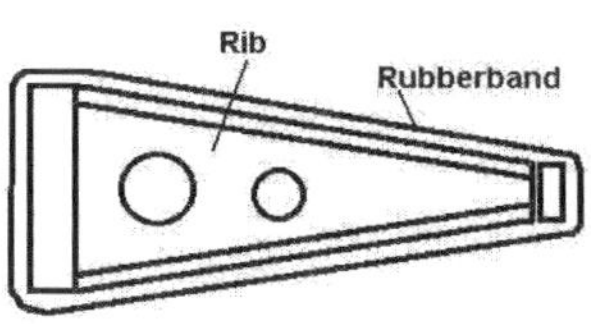

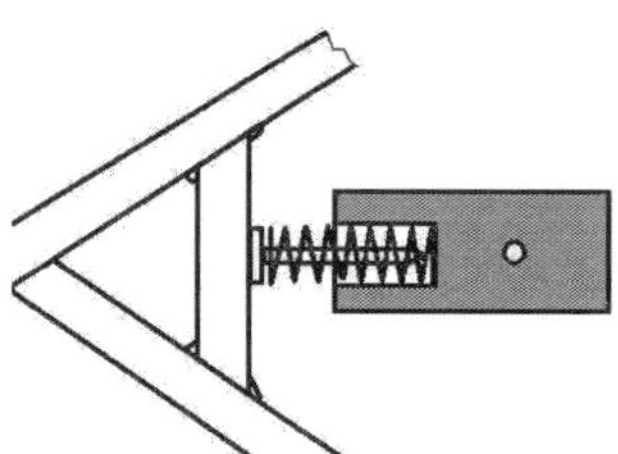

Elastic bands or rubber bands work well for holding ribs or a trailing edge in place.

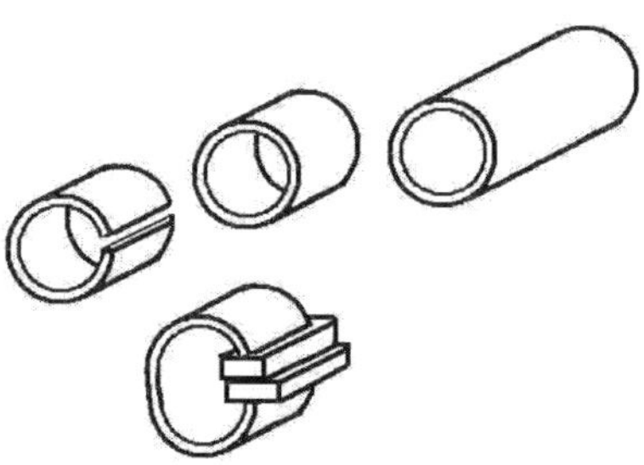

Clamps

Clothes pins work very well as clamps and are dirt cheap when compared with buying large numbers of clamps. Their only limitation is that they can only open so far.

Another cheap source for clamps is making them out of PVC (plastic) pipe. Cut off a section of pipe 1/4 inch or longer, then slit that ring. The opening in the ring of pipe can then be spread and the clamp slipped over the work. A couple of dollars worth of pipe can make more clamps than you could ever want.

Bending Plywood Leading Edges

Wing leading edges made from plywood usually have a tighter bend radius than you can form dry plywood without breaking it. For these applications, the plywood must be preformed. Soak the plywood in the area where it will be bent in hot water until is has become more pliable. Then it has to be fixed in a jig which will hold it in the desired shape until it has completely dried. The jig does not need to have the exact shape of the leading edge, but the smallest radius should be close. For a partial skin, a jig could look like the setup in the left picture. For a complete leading edge, build a jig as shown in the right picture. The tube should be a little longer than the plywood and have approximately the radius of your leading edge. Pull a rope or strap through it. Attach weight to the rope to pull down on the tube or use a ratchet type strap.

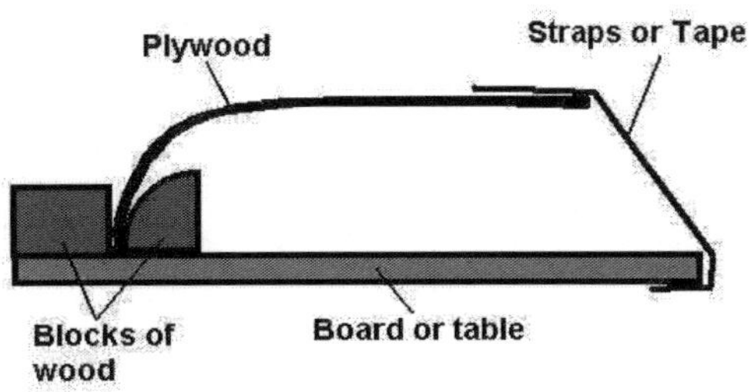

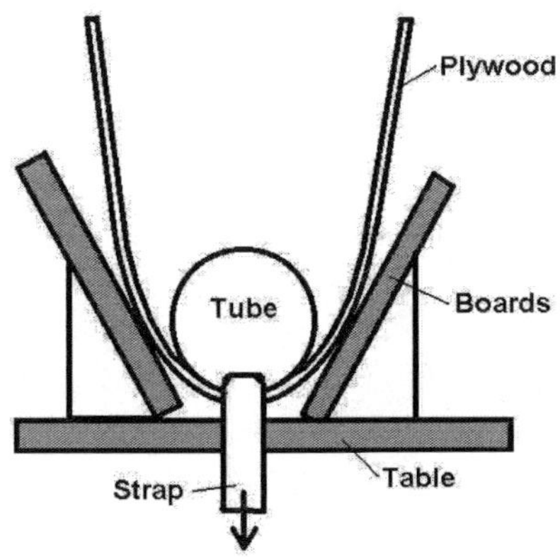

Wood Wing-Fuselage Fillet

You do not have to make your wing-fuselage fillets from composite to get nice curves. They can also be made from wood, more suitable for a wood airplane. Make ribs with the shape of the fillet in the different locations along the chord and glue them in place. The curves vary, so you may need a lot of plywood strips to cover the gap. Prepare the sheets and strips of plywood, edges scarfed, to be bonded on the ribs between the wing and fuselage.

The sheets are painted with epoxy to preserve them except the areas used for gluing.

The completed fillet is sanded until the shape and the joints are smooth. The pictures show the completed fillet and the prepared sheets laid out.

Adjustable Wing Rib Jig

A jig for assembling ribs consists of blocks and supports to position the pieces for gluing. To minimize bond gaps, it is important that the parts fit together well. A bond gap which is too large can substantially weaken the structure. To reduce the risk of large bond gaps, an adjustable rib jig can easily be built. The setup is shown in the following picture. The positioning blocks are made from plastic, Teflon or nylon, which prevents the glue from sticking to the jig.

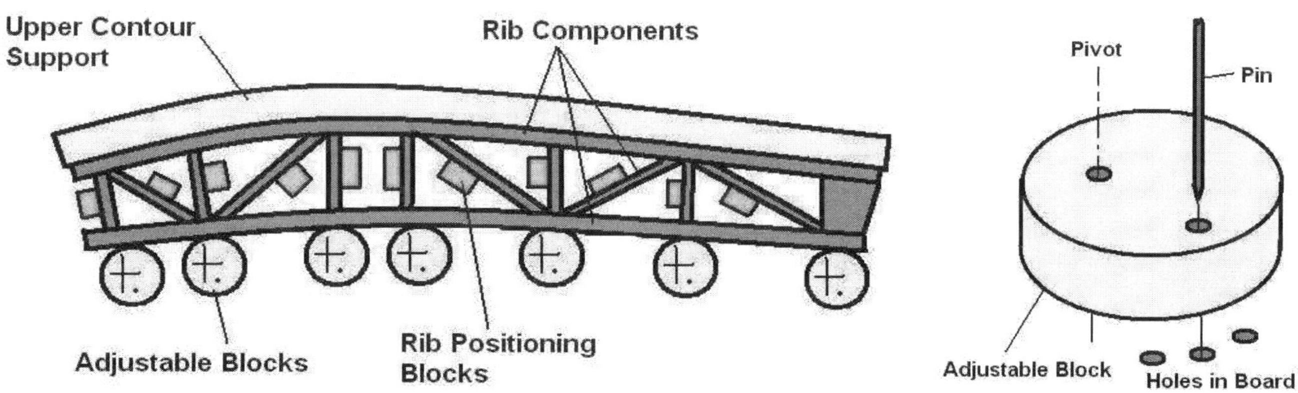

The upper contour support is fixed and must be shaped per the rib plot drawing. The lower side supports are adjustable. They consist of round blocks with an off-center pivot (nail). They are rotated until they press firmly against the bottom rib strap and are fixed in this position with a pin. This kind of rib jig is extensively used by a company building replica World War 1 aircraft (www.thevintageaviator.com). They also have other detailed, interesting construction pictures, mainly wood airplane components, on their web site.

Flight Controls

Balancing Control Surfaces

Most control surfaces require mass balance to eliminate flutter. The designer should have specified a hinge moment or at least an amount of mass to be used as the counterweight. Since the weight and unbalanced moment of the control surface varies between airplanes, only the hinge moment lets you determine the correct amount of mass balance. Control surfaces should be built as light as possible to reduce the amount of mass balance needed.

1. Determine the hinge moment of the finished control surface without any mass balance.
2. Compare the hinge moment with the requirement from the designer.
3. If the hinge moment you measured is larger than the required hinge moment, add mass balance in front of the hinge line.
4. If the measured hinge moment is smaller than the required hinge moment, you can leave it as is or remove some mass balance to save weight.
5. Once you have made a change, either by adding mass balance or if you made a modification to the control surface, re-measure the hinge moment to ensure it matches the required hinge moment.

The normal method of balancing control surfaces consists of supporting the control surface at its hinges with one of the following methods. In method 1, a string is tied to the hinge bolts. If this does not work because the hinge is not accessible from that side, use method 2. The main thing is to have as little friction at the hinges as possible to be able to measure the actual moment. This often precludes measuring it with the control surface installed on the airplane. The hinge moment is the weight measured at the trailing edge times the distance to the hinge. A control surface that is 100% balanced does not have any weight on the trailing edge. "Underbalanced" means the control surface is trailing edge heavy. Overbalancing is undesirable.

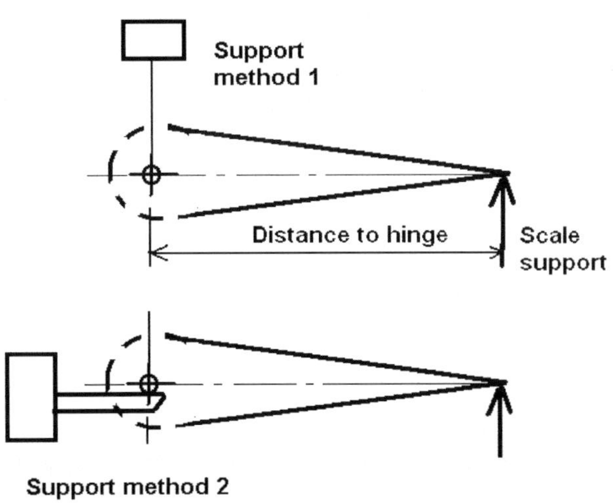

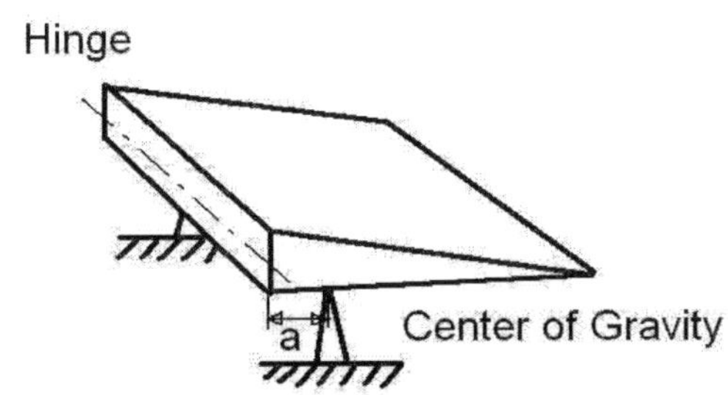

A third method for support to determine the hinge moment is shown in the right figure. Determine the weight of the control surface. Then support it at its ends in its center of gravity. The control surface should just balance on the supports without tipping either way (with a little help from you to keep it from falling down). In this case the hinge moment is the weight of the part times the distance a.

Control Surface Leading Edge

In the case you build a composite control surface as part of the wing (or tail) and cut it from the completed section, it may need a rounded leading added to complete it. One method is adding blocks of foam to the control surface and to shape them by sanding. To make it easier to sand them to the proper shape, use templates attached to the ends. Once you have the desired shape, lay several plies of the same material as the control surface skin over the foam to create a stiff structure.

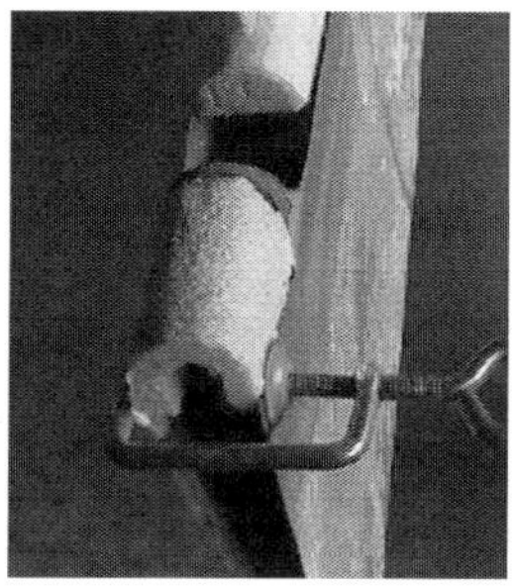
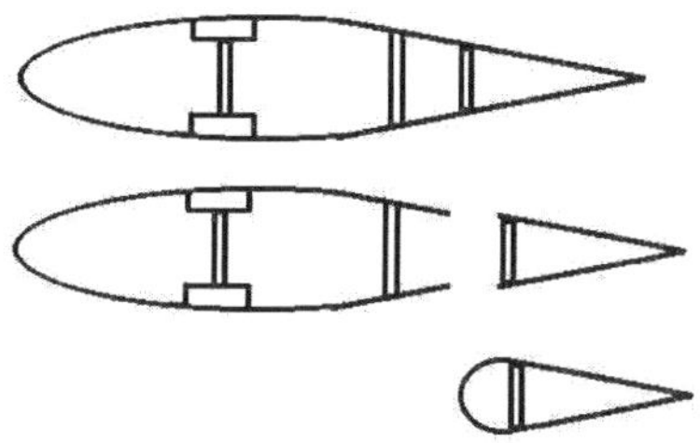

Lightweight Rudder Cables

Traditionally, rudder cables are made from steel. There is no reason why you could not use a lighter material, such as Spectra or Kevlar, if you design for it. Spectra ropes are used extensively on boats because of this material's exceptional strength. A Spectra rope of 1/4 inch diameter can easily replace a steel cable of 3/16 inch diameter and has less than half the weight of the steel cable. To attach it, simply use a knot. The knot should be tight against the bolt it attaches to. The use of a thimble is not necessary. If you plan on using a thimble, test it to the maximum load expected in service. I tested a bowline knot without a thimble on the 1/4 inch rope to 1000 lbs and it held easily. The rope itself had a breaking strength of >2000 lbs. If the knot is not tight, it will slip until it is tight. In order to avoid excessive slipping of the knots, the cable assembly with all knots should be pre-stretched to >300 lbs or whatever your expected load is. This tightens the knots. A higher load should not result in more slipping, but this depends on the diameter and rope used. The rope itself stretches very little. If you use rope, allow for penetrations through the structure which are large enough to let a knotted end pass through. ¼ inch Spectra rope is readily available, smaller sizes are not. This picture shows a 1/4 inch Spectra cable after it had been tested. The knot had tightened so much that it was impossible to open it. It had to be cut to remove it from the test fixture.

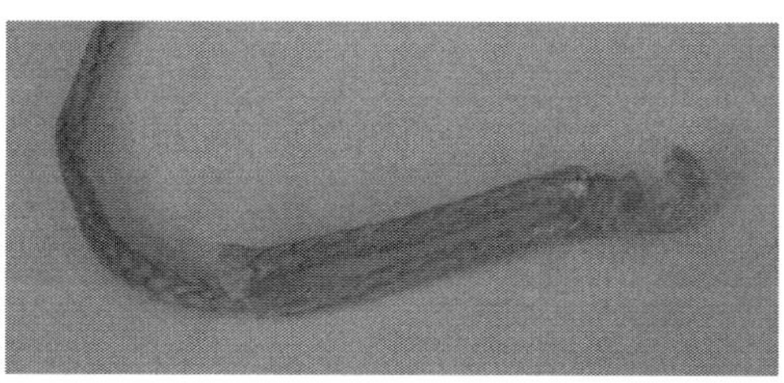
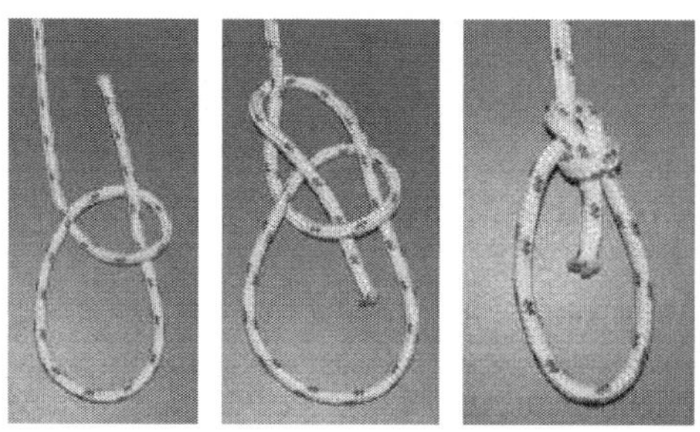

Bowline Knot

Flight Control Linear Bearings

Long control rods may need support in one or several places to keep them from buckling under compression loads. The unsupported length depends on the diameter, thickness and material of the control tube and on the loads. You can make simple, inexpensive, low friction linear bearings yourself.

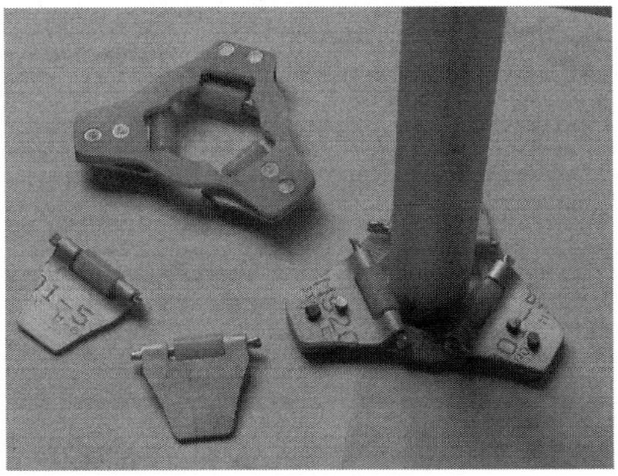

If your airplane is composite, make the base plate from thick, flat composite material so the bearing can be bonded in place (as shown). In a metal airplane, use aluminum for the base plate. Cut a piano hinge in segments as shown. Buy plastic spacers (from McMaster Carr for example) which will fit in the cutout of the piano hinge. I used a brass rod as the pin and deformed it slightly at the ends in a vise to keep it from falling out. Rivet the segments to the base plate (countersunk rivets) so that they support the control tube from three sides. The control tube should just have a little play to move freely in the bearing even if it is not perfectly perpendicular to the surface. The bearing can then be bolted, riveted or bonded to the airframe. Choose a suitable shape for the base plate to work with your attachment method.

A different way is to use sliding bearings (bushing) from a low friction material like Teflon, Rulon etc. They will add a little more friction to the system than the described roller bearing.

Sealing Control Rod Penetrations

Where the aileron control rod exits the fuselage there needs to be a hole large enough to prevent the control rod from chafing. Unfortunately this hole will also let air flow through, either from the outside into the cabin, creating a cold draught, or the other way creating drag. To seal the hole, you can use a bellows. Make sure it allows enough travel of the control rod. The picture shows the ideal situation, with a bellows of the correct size. The bellows is attached to the control rod with a clamp and is glued or screwed to the structure.

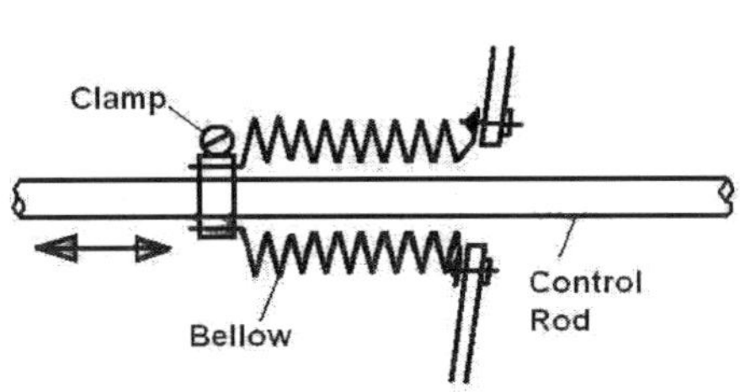

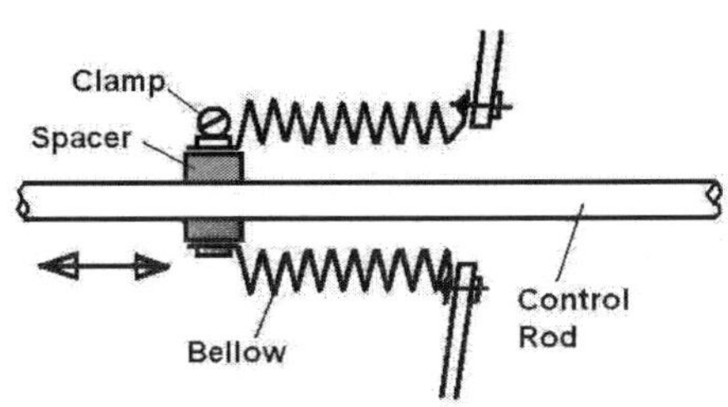

A bellows of just the right size may not be available for your application, but you can adapt it. If the hole is too large, fabricate a spacer from thick plywood. Drill a hole of the size of the control rod and shape the outer diameter to fit the bellows. Glue the spacer to the control rod and clamp the bellows to it. The bellow should be large enough to not chafe on the control rod. If the outer diameter is large, it may interfere with other structure. In this case you could make the hole in the spacer offset to one side. Neoprene is a suitable bellows material with good chemical resistance and suitable to a wide temperature range.

 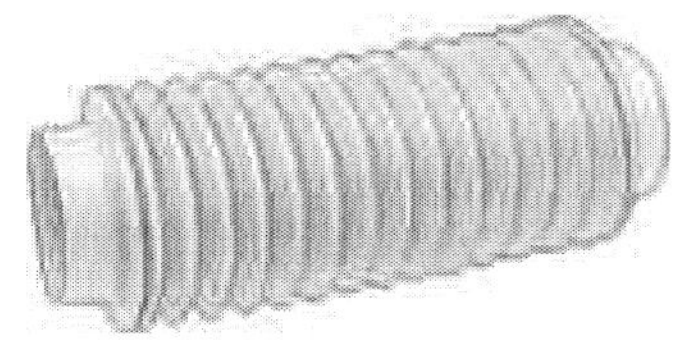

A different method creates slightly more friction. A close fitting funnel is built up from composite around the hole. A piece of firesleeve is then pulled over the funnel and is secured with a clamp or tie-wrap. I would be a bit concerned about the long term effects of the glass fibers from the firesleeve rubbing against the soft aluminum. Maybe a piece of fur or felt would be better?

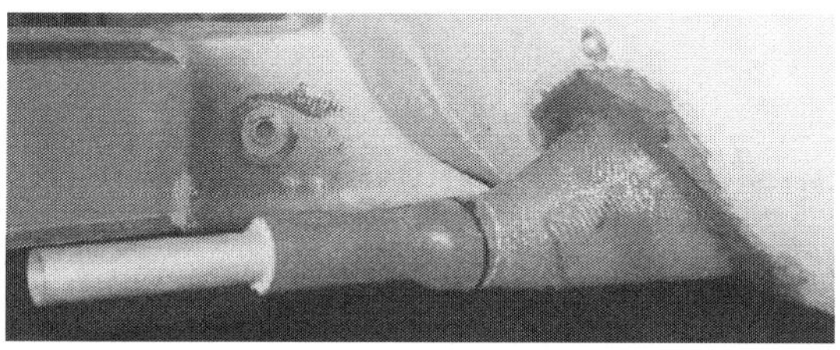

Whatever method you choose to use, think it through what would happen if your design would come loose. Can it possibly block your controls? If the answer is yes, don't use it.

Estimate Control Surface Mass Balance Weight Increment due to Paint

Most control surfaces require mass balance to ensure there is no flutter in flight. If you balance a surface before it is painted, you will have to repeat the process after paint if you do not take the weight of the paint into account. In order to do that, you need to estimate the paint thickness and the surface you are covering to come up with the weight and CG of the paint. Typically the paint thickness is 0.008 inch to 0.015 inch thick (0.2-0.4 mm). The same is valid for primer or filler, you have to estimate that as well. To save weight, use the least amount of filler necessary because it will need to be compensated for with extra mass balance weight.

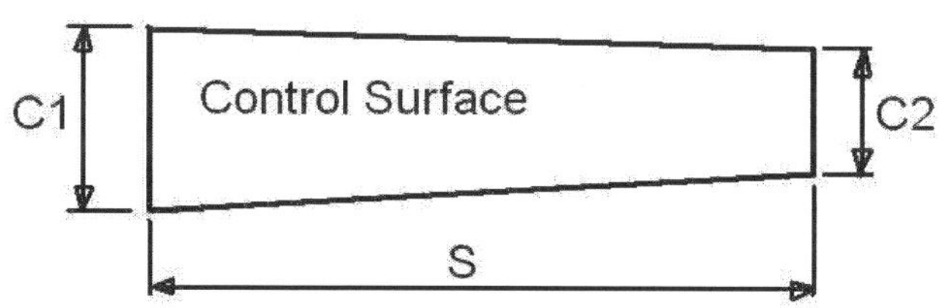

The surface A of the control surface is A = (C1 + C2)/2 * S. Multiply this by two because there are two painted sides. The weight of the paint is the surface area times the thickness times its specific gravity. You can assume about 0.04 lb/in^3 (if you use inches) or 1100 kg/m^3 (if you use metric units) for the paint if you do not know what it actually weighs. When you balance the unpainted control surface, simulate the paint by placing a cup of water of the weight you calculated for the paint in the centroid of the control surface (roughly in the center). The centroid is the center of gravity of the control surface skin and can be found by balancing the skin over a piece of angle iron in the long and short dimension. Mark it so that it is not obscured by priming and filling or better note its position on a picture or drawing of the piece.

Flight Control System Stiffness

It is important to build the flight control system with sufficient stiffness. If for example you clamp the stick in a fixed position and apply 2-5 lb (1-2 kg) of force to the control surface, the trailing edge should not move more than 1/8 inch (3 mm) total. Any more flexing or play can degrade the flight characteristics and in the worst case cause dangerous oscillations or flutter.

Freeplay or flexing can be the result of loose bearings, flexing bellcranks, bending control rods or lack of stiffness of the bellcrank supports. It may be aggravated by an unsuitable gearing ratio. By this I mean the stick deflection ratio to control surface deflection. You would have an unsuitable gearing ratio if your stick moved only one inch to move the elevator through 30º of deflection. The same concept applies to other components in the control system as well. Small movement on one end geared to a large movement on the control surface amplifies any freeplay.

To avoid problems, look at your control system components while you are building the airplane from the stiffness point of view. This is the time to fix or improve it without too much trouble. Once the airplane is assembled, it may be a lot harder to get to the respective components.

Flight Control System Bearings

Bearings in control systems have two jobs: to support a control surface like an aileron or elevator and allow them to move with the least amount of friction. How much friction is acceptable depends on the airplane. It usually is a tradeoff between cost and complexity. The simplest bearings are bushings. Usually the hinge or bolt is steel for structural reasons. The bearing material in combination with the hinge material determines the load rating. The larger the bearing surface, the higher is the load it can handle. Keep in mind that normally in flight, there always is a load on the control surface, stressing the bearings and potentially increasing the friction. The following table lists advantages and disadvantages of the different styles of straight bearings.

Bearing Style	Load Rating	Friction	Cost	Maintenance
Ball Bearing	Low	Low	High	None if sealed
Roller Bearing	Medium	Low	High	Some if not sealed
Brass Bushing	High	Medium	Medium	Lubrication lowers friction
Teflon or Rulon Bushing	Low	Medium	Medium	None, but will wear quicker
Steel Bushing	Highest	High	Low	Requires lubrication, grease nipple

The picture shows a selection of different bearings. There are of course other considerations as well, such as size. Ball bearings may require more space than a simple bushing, but if low friction is the objective, that is the way to go. Bearings are also used in the control linkage. In many places some angular movement is required. There are only two types of bearings which will allow this: rod ends and self-aligning bearings. Both require lubrication and are open to contamination. The tolerances of rod ends vary widely, even from the same supplier. Some are really stiff and some may rotate easily. If you use the stiff ones, the friction in the control system will be high. Over time they will wear in slightly. To loosen a rod end up prior to installation, you can clamp it in a drill press and spin it for a while. I have seen simple hinges made from sheet metal often show worn bolts and enlarged holes and therefore create a lot of play. The strength of such hinges may be adequate for the static condition, but they do not last long under the rotating motion of a control surface. To increase service life, the bearing surface must be increased.

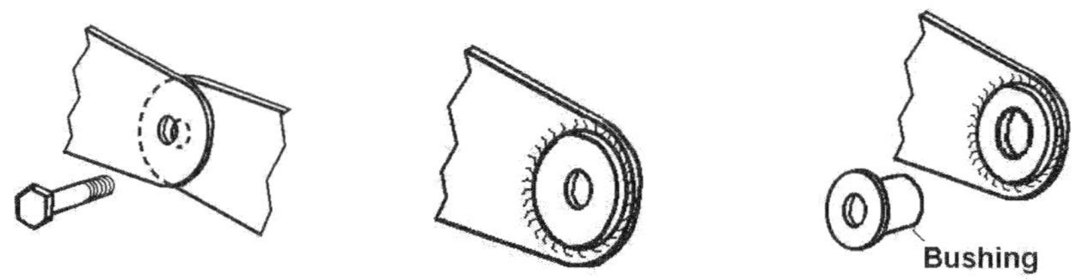

Not good Better Best

This is easy to do by just welding a washer to one or both sides of the hinge (use a bolt to hold the washer in place while welding it). It is even better to use a bushing made from a suitable bearing material such as brass in addition. The bushing should have a press fit to avoid having the bushing rotate in the hinge rather than the bolt in the bushing.

There should not be a need to replace any bushings or bearings during the life of the airplane if they are designed properly. Because of some operating conditions are more severe than others, of course there may be cases where bushings may wear more than expected and need replacement. The width of a properly designed bearing or bushing is suitable for its material. A bushing from softer material will just have to be made wider to result in a lower load on the bearing material to achieve a long life.

Control System Functional Check

Before flying the airplane, perform a very thorough check of the control system. It can easily happen that lines, wires or structure are touching components of the control system. All moving parts should have at least 1/4 inch clearance to everything else. Check that the sticks or yokes do not hit the engine controls in any position. Move the controls through their full range and verify that the controls move in the proper direction and that bellcranks swinging through an arc do not contact anything. In flight, the airplane will flex and marginal clearances could cause binding under load.

Measuring Control Surface Deflections

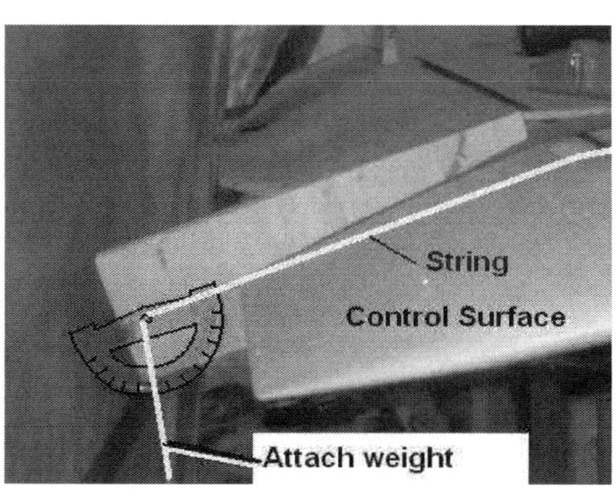

As a final check prior to flying your airplane, you should measure the deflections of all control surfaces and verify they are what the designer prescribed. There are many ways to do this. If you have a digital inclinometer, it is simply a matter of placing it on the control surface in the neutral position, setting it to zero and measure the angles. This does not work with the rudder, unless you can turn the airplane on its side. Another way is shown in this picture. Attach a wood template to the control surface and a small goniometer behind the trailing edge. Attach a string to a screw in the center of the goniometer. Attach a small weight to the string to make it point straight down. With the control surface in the neutral position, rotate the goniometer so that the string is at 90º. Now you can deflect the control surface in both directions and you can read the angle on the scale. The deflection is the difference between the reading and 90º. This still does not work for the rudder.

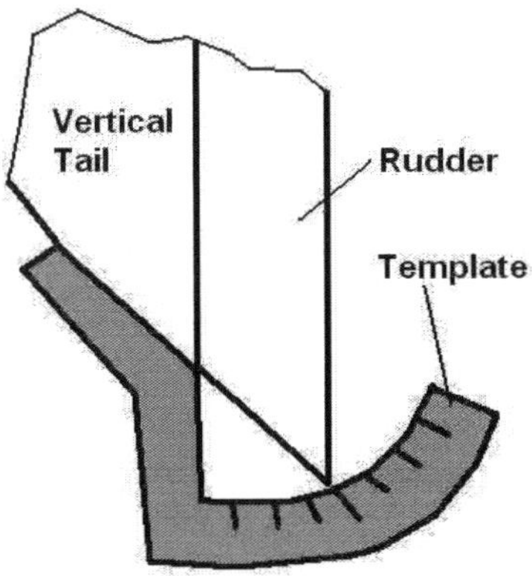

For the rudder you need a template. The template will rest on one side on the vertical stabilizer and have a cutout for the rudder large enough to allow full deflections. Select a location on the rudder suitable for the measurement. This can be the bottom edge or any other section. Mark it on the rudder. Measure the distance from the hinge to the rudder trailing edge C. You should mark the template every 5º and at the full required deflection. The units used here are inches, but you can use any other units as well. The spacing of the degree marks can be calculated easily:

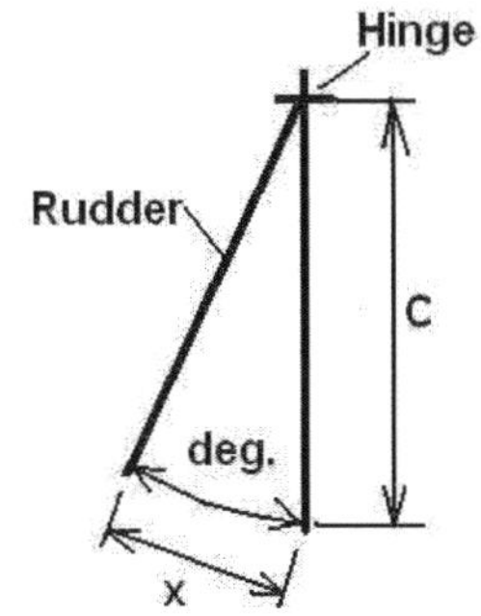

The distance between x degrees = 2 * C * sin (x/2)
Example: C = 7 in, Full deflection = 24º,
Distance to mark on template for full deflection = 2 * 7 *sin (24º/2) = 2.91"

Cable Retainers for Pulleys

Pulleys need a cable retainer to keep the cable from coming off the pulley, should the cable become slack momentarily. You must assure that the retainer cannot rotate and possibly start rubbing on the cable. The two legs of the retainer in the following picture are riveted together on a separate tab. One method to prevent them from rotating is to add a longer leg on one side which allows it to be clamped to other support structure as in the left and center picture below. In the right picture, the retainer is pinned to the pulley support bracket.

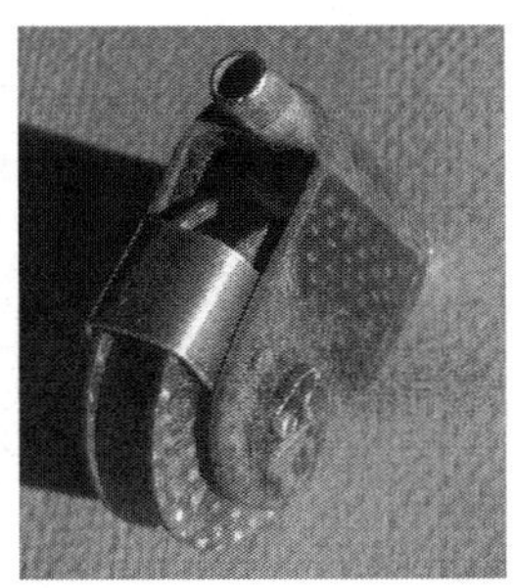

Large washers on either side of the pulley are used to prevent total failure should the bearings somehow come loose.

Control Cable Fairlead

Ideally, control cables are routed in straight lines. If it happens that a cable might run too close to structure and would rub on it, or it needs to be redirected a small amount, a fix is needed. Rather than use an expensive pulley, you can use a fairlead to guide the cable as long as the angle which the cable is deflected is small enough. Fairleads should also be used if a cable is unsupported over several feet, might develop slack and rub on something. A fairlead can be made from a plastic bushing and an appropriate support. Rudder cables may also be guided in conduits. Plastic tubing such as nylon or polyethylene are suitable and cause little extra friction.

Control System Stops

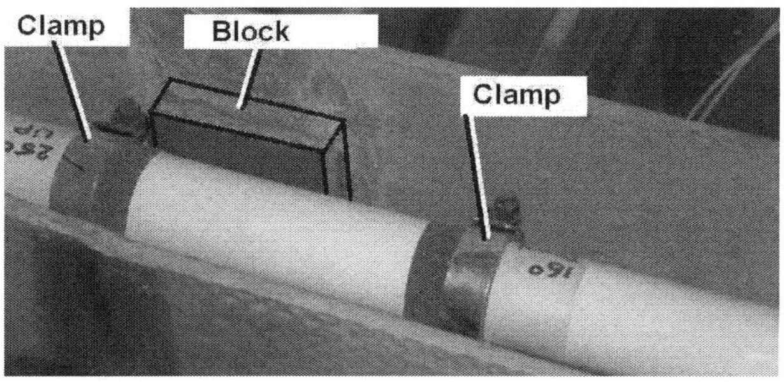

All control surfaces should have positive "stops" to limit their travel. To prevent damage to the control system if a wind gust on the ground slams the control surface around, sturdy stops should be located right at the control surface. Stops at the stick should be installed in addition to the stops at the control surface to prevent an overly eager pilot from damaging the system during a preflight check. Stops can be welded on, bolted on or bonded on depending on the construction method of the airplane. At least one of the two stops should be adjustable, ideally both. The stops on the stick and on the control surface should be reached at the same time. The stops at the control surface often consist of a screw which contacts a hard point on the hinge. The screw is secured with a jam nut. A different arrangement can be used in the cockpit. In this example, worm drive clamps have had flat faces brazed onto the worm gear. They are clamped around the push rod. A block of suitable material (phenolic for example) is glued and or bolted to the sidewall. The clamps are very easily adjusted to give required travel.

Simple Cable-actuated Trim Tab

There are many ways to build trim tabs and actuate them. Some require electric motors and may be quite complex. Two concepts are presented here which belong to the opposite end of the spectrum because they are very simple. The first one consists of a sheet metal tab attached with a piano hinge to the trailing edge of the elevator. A solid wire push-pull cable actuates it. The actual installation is shown in the photo on the left, the schematic is shown on the right. The cable guide is external and needs to have enough slack to allow for full elevator deflections.

The main disadvantage is that there is no redundancy of the actuation mechanism. If the cable breaks or becomes disconnected, the trim tab is free to move. This is an excellent way to induce elevator flutter which is likely going to lead to the destruction of the airplane. One way to avoid this is to provide a second actuator, friction device or damper which would prevent free movement of the tab in case of the failure of the primary actuator.

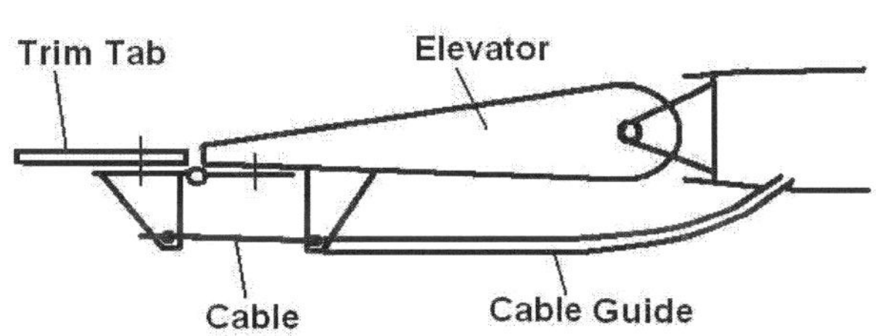

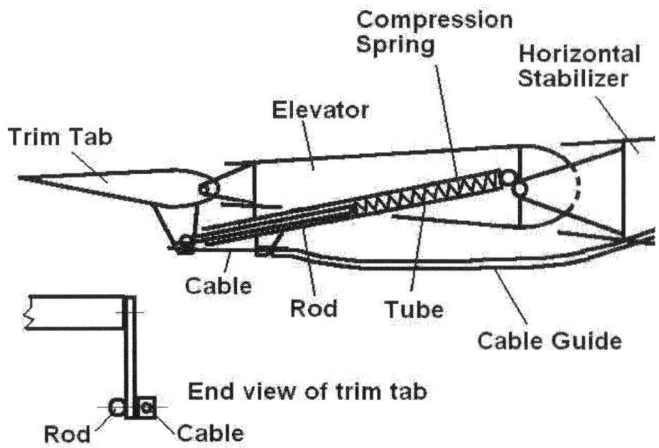

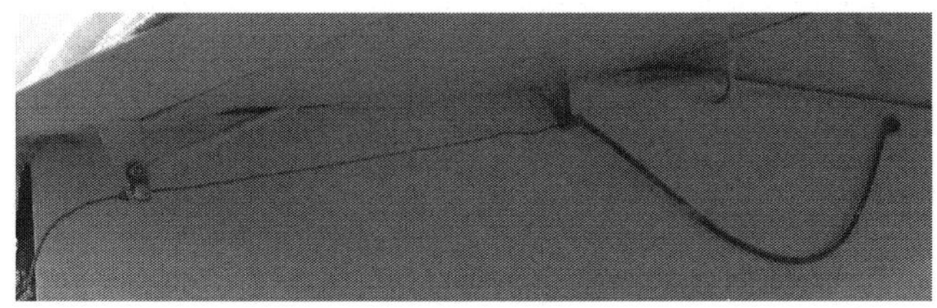

One example which has the required redundancy in a very simple manner is shown in this picture. The trim tab is operated by a cable which only has to provide a force in tension to pull the trim tab down. A rod in a tube with a compression spring provides the force to push the trim tab up. If either one fails, the other one provides force in one direction to force the trim tab to the stop on one end. This means the pilot will have to compensate for a mis-trimmed airplane, but at least there is no risk of flutter. The cable guide is external as well and requires enough slack to allow full elevator deflections.

Casting Lead Counterweights

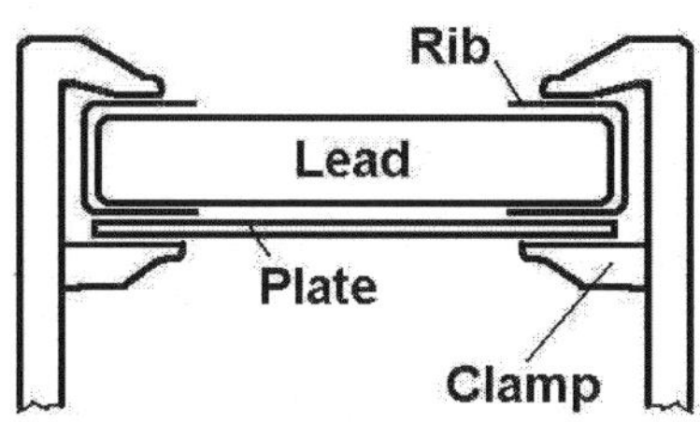

A welded tube structure control surface allows a unique way of attaching the counterweight: casting it in place. A piece of aluminum plate is clamped to the bottom of the ribs with some ribbon caulk as a sealer. The molten lead is then poured into the compartment created by the ribs and plate. This way, the lead is contained in the ribs. It can later be drilled out to achieve the desired hinge moment.

Instead of melting and casting lead directly, in some cases you may have the option to use lead shot mixed with resin to fill up the space provided for the counterweight. This avoids high temperatures and can be used with composite structures. Or a chunk of lead can be roughly shaped to fit its designated pocket and be bonded in with cotton flox. Be careful when working with lead, it is highly toxic. Wear gloves when touching it. Its use should be minimized. If possible use alternatives like brass or steel.

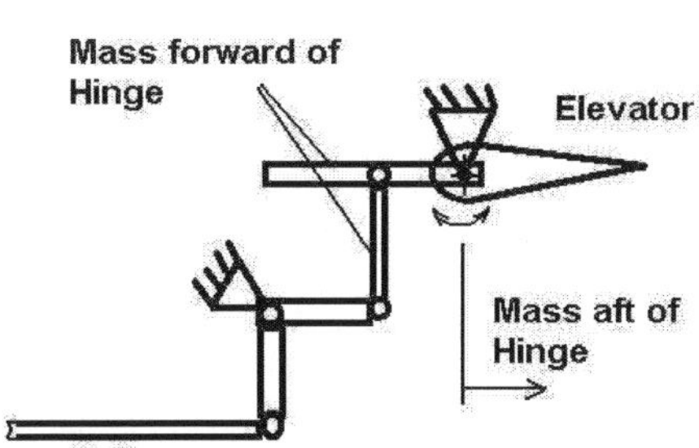

Internal Mass Balance

Not all mass balance has to be external or directly part of the control surface. In some cases it is possible to use the geometry of the control system to take over the function of mass balance, at least for the elevator. This sketch shows the geometry. The control linkage inside the fuselage, which is forward of the elevator hinge, balances the weight of the elevator. Since the arm can be made fairly long and still be within the fuselage, the mass can be much lower than on a short arm and result in the same hinge moment.

Electrical Systems

Electrical Grounding

If you are building a non-metal airplane, you should ground all metal components in your airplane. In a metal airplane you can be pretty sure that there is very little electrical resistance between the different components. The only ground strap needed in the latter case is typically the one from the engine to the airframe because rubber isolator engine shock mounts are non-conductive. It is different in a wood or composite airframe where metal components are usually isolated from each other. To avoid sparks from static discharge when you touch a metal component, connect all metal components with ground wires. Sparks can be an ignition source, and must especially be avoided if fuel fumes or oxygen are in the vicinity. Other effects from poorly grounded components may result in radio interference and buildup of electrical charges when flying in certain weather conditions.

Wire Splicing Method

In this paragraph it will be demonstrated how to make wiring splices that are as good and strong as no splice at all. Of course this is not the only wire splicing method, but it is one that is simple and works well. This splicing method uses "parallel connectors", available from Waytek, a company which does have a lot of very handy wiring products. They also have various waterproof connectors, as well as the "dual wall" heatshrink you will need to complete these splices. The dual wall heatshrink has a heavier wall thickness and it has an inside wall made of hot glue. After you have shrunk it, whatever is inside is completely protected and encapsulated in plastic. The parallel connectors come in 3 sizes - the smallest is p/n 30070, medium is p/n 31070 and large is 32070. First, cut back the insulation 3/8 inch on the wires to be spliced. Put the parallel connector onto one wire, making sure that a bit of stripped wire is exposed on either end. To avoid ticking yourself off and having to redo a good splice, **make sure you put the heatshrink onto one of the wires now**. Note the heatshrink on the right side of the picture below (center). Use dual wall heatshrink, cut to about 1.5 inch length. You want to make sure it will go well beyond the splice in both directions. Waytek part numbers for this stuff include 22202, 22203, 22204, etc, depending on size. Slide the other wire in from the other end of the connector, making sure that a wire end is exposed at each end of the connector. Make sure that all individual wire strands go through the connector, and do not get hung up at the entrance. Use the smallest connector that you can put all strands of both wires into. You can also splice multiple wires into either end, as long as they all will fit. Use the smallest heatshrink that will fit over the connector.

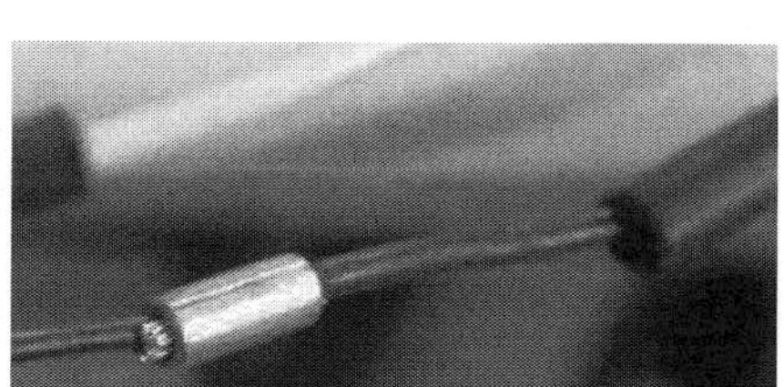
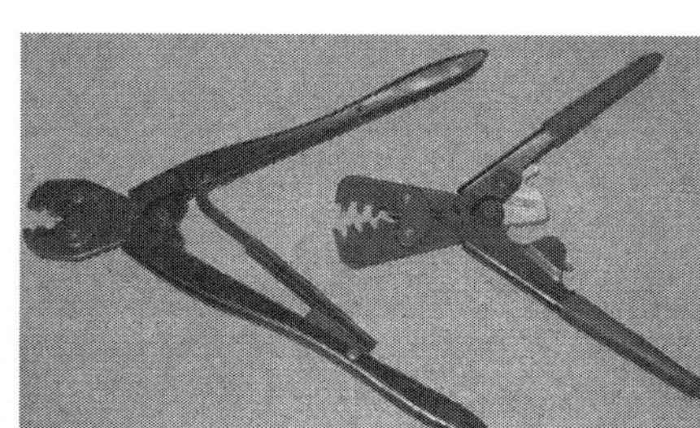

These are the crimpers you can use for this. On the right is the one Waytek recommends for it. It is their p/n 464 Metri-Pack (waterproof connectors) crimping tool. It does a fairly good job of crimping the parallel connectors. Better is the AMP tool (p/n 69324) on the left, if you can find one. You could also use most any PIDG crimping tool.

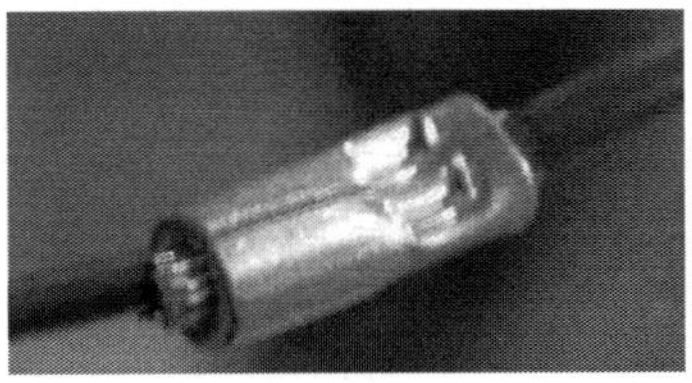

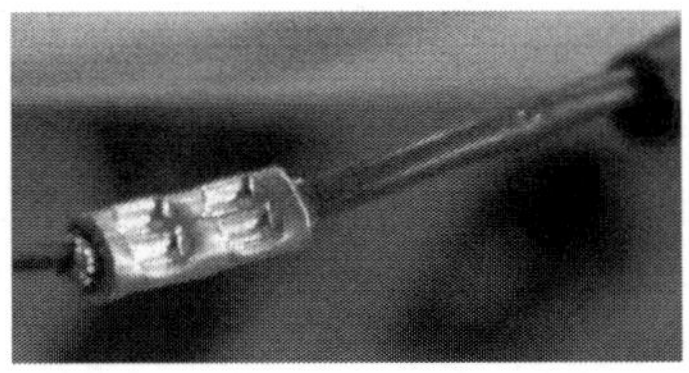

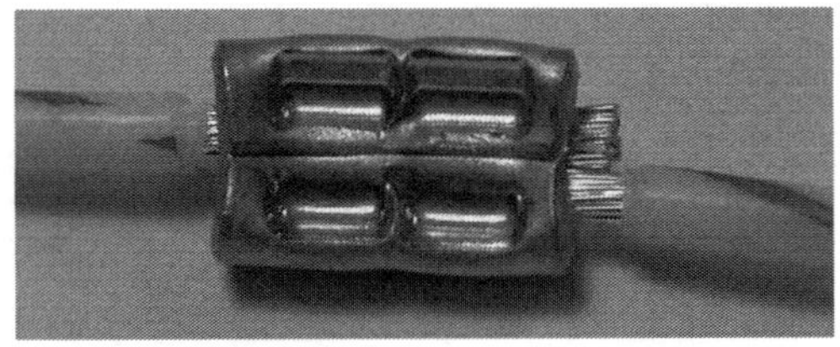

Align the splice with the seam up so you can see it. Center the crimper male jaws over the seam. Position the crimper axially so the crimp will not touch either the end or the center of the connector. Before doing this, make sure the dual wall heatshrink is in place on one of the wires. Then crimp it again on the other end, again avoiding the very end of the connector and the other crimp. The center and right pictures show what the final crimping will look like. Actually, the bottom wire should have been a bit more to the left, so that the end would be showing. Then slide the dual wall heatshrink down, center it over the connector, and heat it thoroughly. You can see the plastic glue seeping out from the right end. The heatshrink needs to be about 3 times as long as the splice it is covering. The dual wall heatshrink is mostly for protecting the splice against corrosion, fraying and flexing, as well as the obvious insulating.

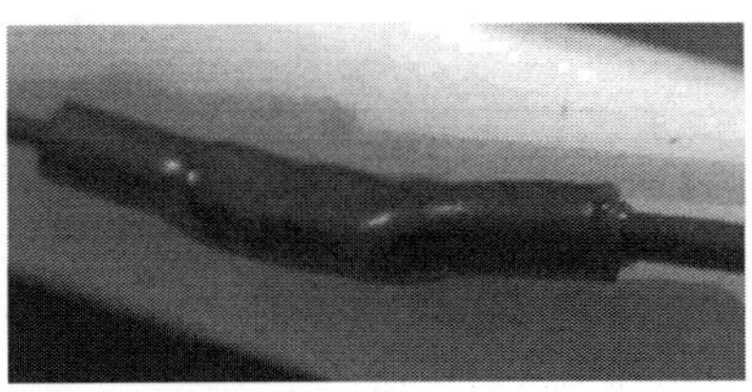

Make the first splice you make with this method your test piece and try to pull it apart after is all finished. It should be very difficult to break this splice.

Ground Power Connection

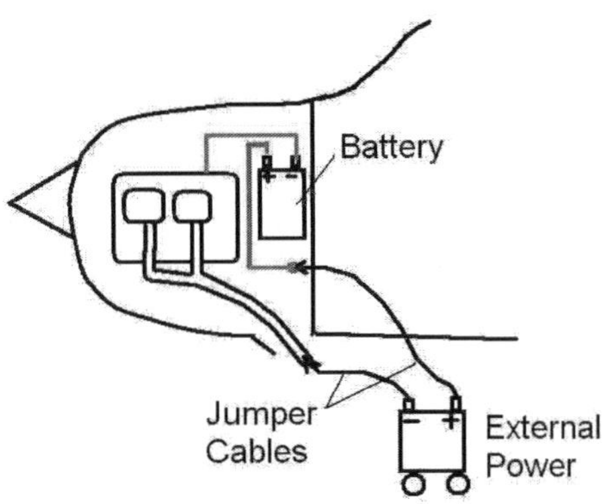

If you live in a cold area where the battery can easily get worn down starting, make a jumper attachment through the cowling or fuselage. It is best to use a ground power plug which protects the terminals. A simpler arrangement, which lets you use jumper cables without having direct access to the battery, makes use of the exhaust as the return path / ground. The power terminal can be as simple as a bolt hard-wired with a heavy gauge wire to the + battery terminal. Keep in mind that this bolt must never be shorted to ground or there will be some impressive sparks and soon a dead battery. External jumper cables can then be connected to that bolt (+) and the exhaust (ground) to charge the battery or start the engine.

How to Lace a Wiring Harness

Lacing a wire harness means tying bundles of wires together. You need lacing tape (from Aircraft Spruce for example). Typically it is black in color although in the pictures shown here it is white. Tie your wire bundles every couple inches using a **clove hitch knot**. Keep the lacing tape flat and not twisted. To start the knot, wrap a piece of it loosely around the bundle as shown here. You can see a completed wire splice in the right picture.

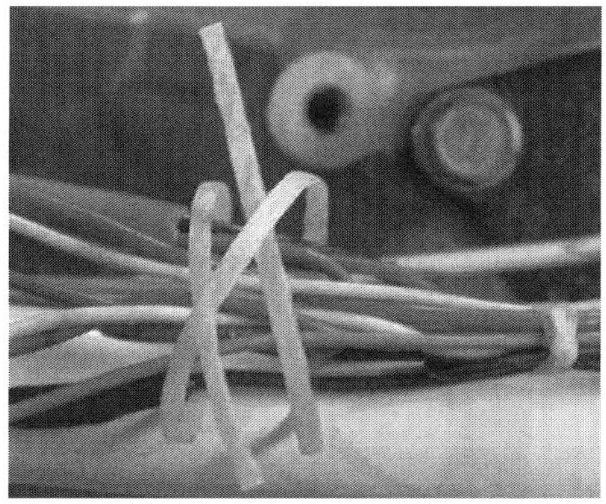

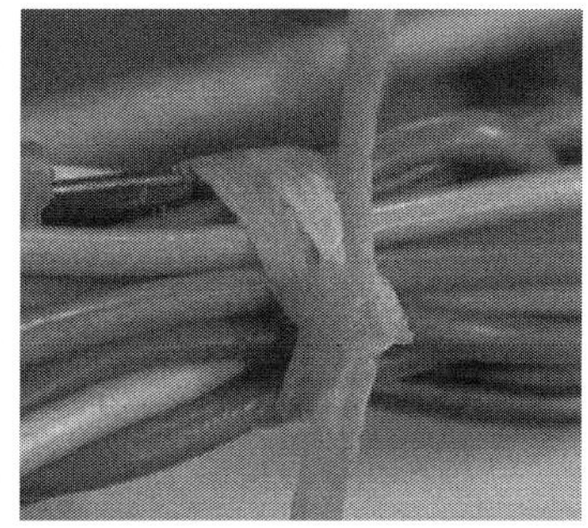

Bring both ends through the loop in opposite directions and pull it snug. Tie it off in a square knot - remember “left over right, right over left”. Snip off all but 1/8 inch-1/4 inch of the ends, then press the ends down into the knot. The wax on the lacing will hold the ends in place. Wire cutters do not work well on this lacing tape. Small sharp scissors work much better.

This picture shows a partially completed harness, with branches as needed. The light colored thick wire is kept separate because it is the starter wire, carrying lots of current. Keep high current or high voltage wires separate so they do not introduce problems into sensitive circuits. As you know, electrical currents flowing through a conductor also create magnetic fields. To reduce their influence on the compass, you should twist wires with currents flowing in opposite directions (positive and negative). This way, the magnetic fields mostly cancel each other out.

Marker Beacon, Transponder and NAV Antennas

If you have an airplane built from a non-conductive material, you can install antennas internally. Instead of using antennas designed to be installed externally, in many cases a copper strip will do the same job. For a home-made marker beacon antenna all you need is a copper strip about one inch wide and 12-15 inches long with a coax cable soldered to the center. Glue it spanwise to the inside of the fuselage skin in a convenient location (left picture). I found that this works as well or better than a store-bought antenna. The center picture shows a NAV antenna made from a copper strip and bonded to a lower wing skin. It has to be installed horizontally and a wing is the most suitable place for it, as long as the skin is not conductive. A transponder antenna needs a ground plane, whether it is installed internally or externally. Fortunately the ground plane only needs to be fairly small (right picture).

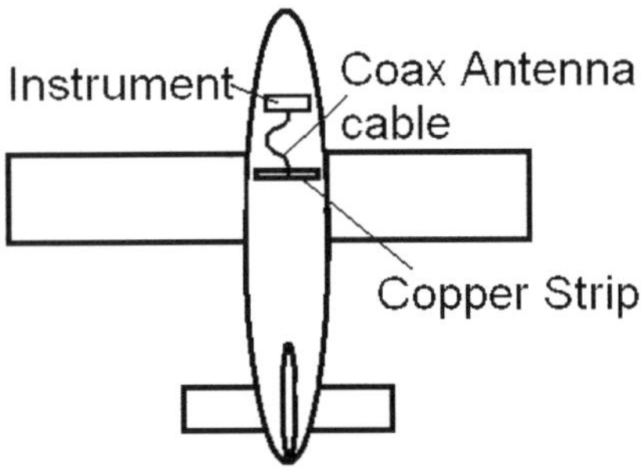

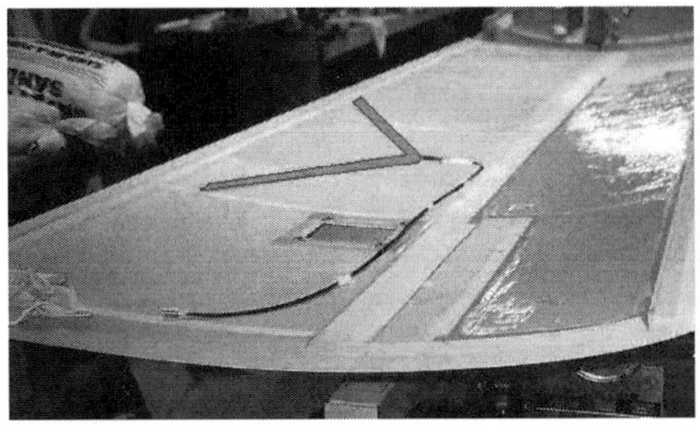
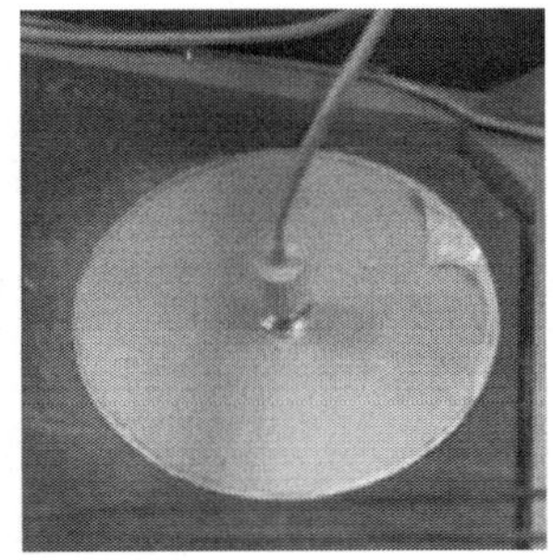

Internal COM Antennas

COM antennas can be a problem because of the length requirement and vertical installation. COM antennas are vertically polarized and therefore require a vertical space of about 46 inches for a half wave dipole. For small airplanes, the antenna has to be shortened a bit. If carbon spar caps, control cables, push rods or wires have the same alignment as the antenna, the antenna should not be mounted nearby because practically no signal will be able to get past the long conductors that will act as reflectors. When installing any antenna, remember that antennas do interfere with each other when installed too close together. Close means one wavelength or closer. One wavelength is about eight feet at VHF frequencies. Less than a quarter wavelength is really bad from both a VSWR and the radiation pattern standpoint.

In a non-conductive airplane, the vertical tail is usually the first choice for the antenna location. One builder describes the installation of a dipole antenna: The antenna is a dipole consisting of a piece of 20 inches long, 1/2 inch wide copper tape laminated into the leading edge of the vertical stabilizer, and a matching length rolled up into an arrow shaft as the bottom leg. Although the connection can be either way, this one has the center conductor going to the upper foil tape, and the shielding was twisted together and soldered to the bottom leg of tape. Some hot glue, micro, and fiberglass tape over the connection prevent the connection from breaking loose. The antenna does not necessarily have to be straight. If your vertical tail is too short, the antenna can be folded into a Z-shape. The antenna should be mounted as high as possible. The center picture shows the "Bob Archer" antenna in a vertical stabilizer. The right picture shows a half-wave dipole COM antenna.

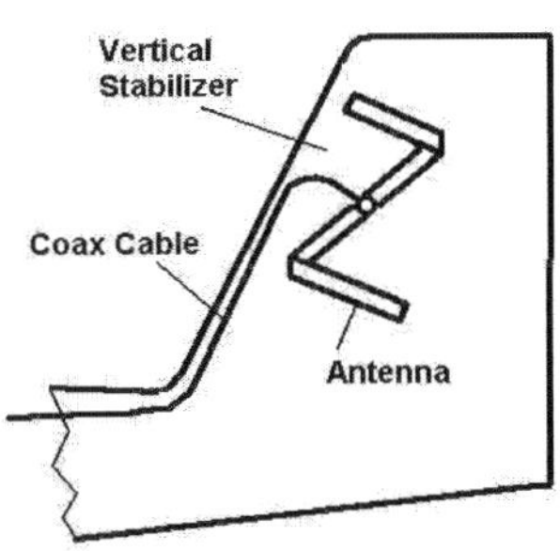

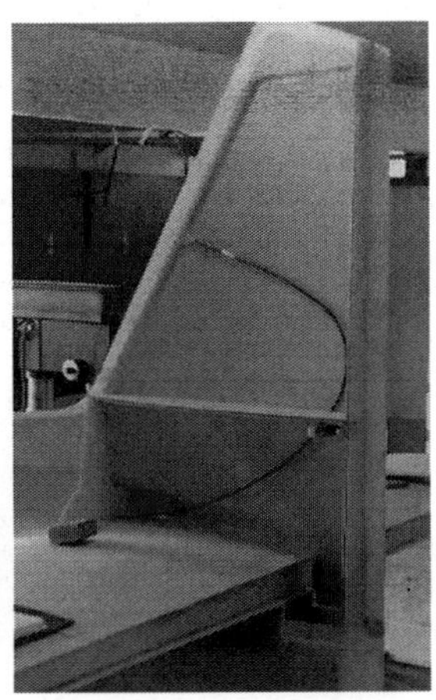

A different style COM antenna is the full wavelength loop antenna. In order to use this antenna, your airplane needs to have a tail cone diameter of at least 31 inches somewhere. A full wavelength loop antenna has a small amount of gain (on the order of a dB or two) over a dipole or quarter-wavelength whip antenna. It has an even larger amount of gain over any antenna that is shorter than a quarter wavelength, such as externally mounted fin type antennas and helically wound antennas. Among the various shapes available for full-wave loops, the circular form factor produces more gain than square, triangle or delta, and rectangular loops. The antenna can be made of any copper wire, copper tape or tubing. In theory, larger diameters (such as 14 gauge wire) should provide a better signal on the edges of the aircraft band, but in practice, this is probably negligible, and you might as well save weight and use the same 20 or 22 gauge Tefzel-insulated wire that you are using for the rest of your wiring.

The wire loop is installed inside the aircraft around the tail cone at the point where the tail cone is approximately 30.75 inches in diameter. Use dabs of epoxy or any other suitable non-conductive material to attach the wire to the inside wall of the fuselage. Since epoxy is a non-conductor, it is not necessary for the wire to be raised or insulated from the surface. The circumference of a circle is $c = \pi * d$ where d is the diameter, and $\pi = 3.14$. To design the antenna to resonate near the center of the aircraft band, at about 125 MHz, a total wire length of 96.5 inches is required. This produces a loop with an approximate diameter of 30.75 inches. The closer the antenna profile is to a perfect circle, the better it works. If you use non-insulated wire, this dimension should be fairly accurate. If you use insulated wire, the insulation can change the resonant frequency and cause slight deviations from the theoretical wire lengths, so you will need to "tune" the antenna after construction. Try to keep the wire as far as possible away from any metal parts, such as fittings, bellcranks, control rods and cables. If you have to, bend the loop into a bit of an odd shape rather than allowing it to get too close to these metal parts. Attaching the loop to the aft face of a seatback bulkhead is not recommended, because there will be people sitting there, and that will interfere with the radio signals as well as producing an unhealthy RF energy field going straight into their backs (the antenna should be at least 3 ft away from a person). Install the antenna wire so the two ends wind up forming a small gap at the 3 o-clock or the 9 o-clock point in the wire loop. It does not matter whether you choose the left or right side. This gap is called the Feed Point, and it is where the coax will be connected. You may wish to leave the ends about a half inch longer than the calculations specify, in order to have some leeway later on when tuning the antenna. Placing the gap halfway up the fuselage side causes the antenna to radiate in vertical polarization. If the gap was at the top or bottom, the antenna will become horizontally polarized with significant loss in signal strength.

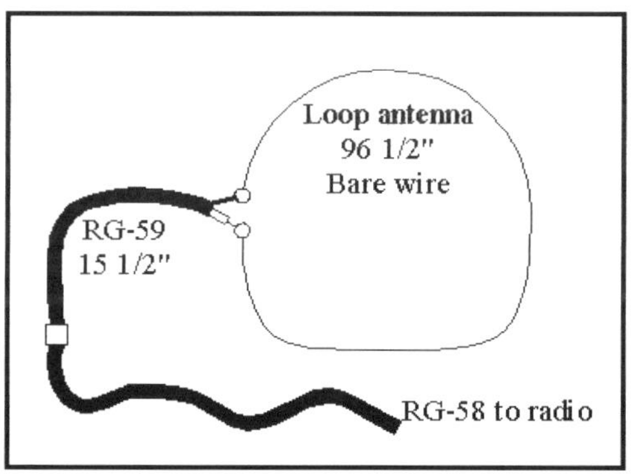

A single full-wavelength loop exhibits an impedance of about 100 Ohms. In order to match this to standard RG-58, RG-400, or RG-142 coax (53 Ohms), a quarter-wavelength matching section is fashioned from a 15.5 inches length of 75 Ohm RG-179, RG-302 or RG-59 coax cable. Attach one end of this coax matching section to the Feed Point described above. Placement of the coax is not critical, but at least the first foot or so should leave the antenna perpendicular to the wire if at all possible.

Antenna gain is always at the expense of directivity. In the case of the loop, the gain is perpendicular to the plane of the loop, which would mean it would have the greatest gain to each side of the aircraft. In theory, there would be a null off the front and back of the airplane. Single loops, such as this, are often used as direction finding antennas, because of this feature.

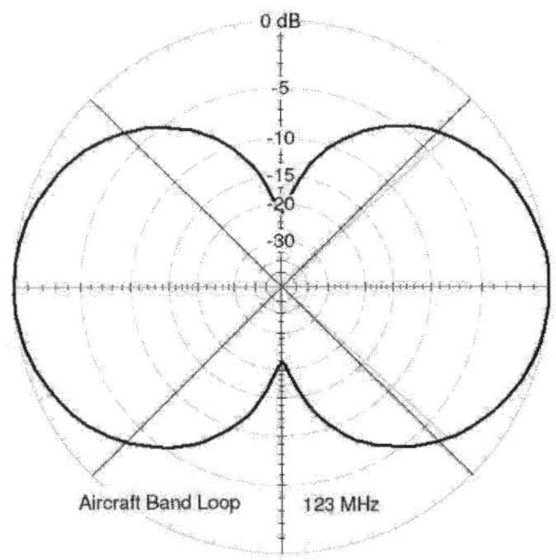

This picture shows the gain calculated for the described loop antenna. You will note the nulls, which would be toward the front and back of the aircraft. Often, that is the most desired direction for radio communication. This null is calculated for free-space and it can be expected that in an aircraft the nulls would be distorted and possibly significantly less than shown on the graph. However, it could pose a problem in at least some cases and the builder should be aware of the potential problem. Antennas are often compromises, and the loop may be a compromise that is still acceptable.

Tuning a COM Antenna

All antenna calculations are estimates and are based on an antenna floating in free space. The theoretical computations cannot possibly account for the impact that other objects will have in the vicinity of the antenna, for example fittings, carbon fiber in the spar, control cables, bellcranks, rods, and so on. So, some minor tuning of every antenna is generally required. Buy an SWR (Standing Wave Ratio) meter that is designed for VHF, or borrow one from a ham who has 2 Meter radios or from an avionics technician. Check your SWR at center of the frequency range most likely to be encountered. If it is less than 1.5 to 1, try the outer ends of the aircraft band. If it less than 1.5:1 everywhere, consider yourself lucky and skip the rest of this discussion on tuning!

If it is over 1.5:1 anywhere, then you might want to do a little work on the antenna to tune it to the right length. If it is over 2:1 then you definitely need to do the following: Tune your transmitter to a frequency about 2 MHz lower than the center frequency and see if the SWR gets better or worse. If it gets better, the antenna is too long. If it gets worse, the antenna is too short. To verify this, tune the opposite direction, that is, 2 MHz above the design frequency and check the SWR again. It should show the opposite results. You can add or remove wire from the antenna at either the coax center conductor or at the braid. It does not matter. If you happen to spot the exact frequency where your SWR dips to its lowest point, you're in luck: just add or subtract 1/2 inch of wire for every Megahertz of offset you need to center it on the desired design frequency. Add wire to bring the center frequency down, and subtract wire to bring the frequency up.

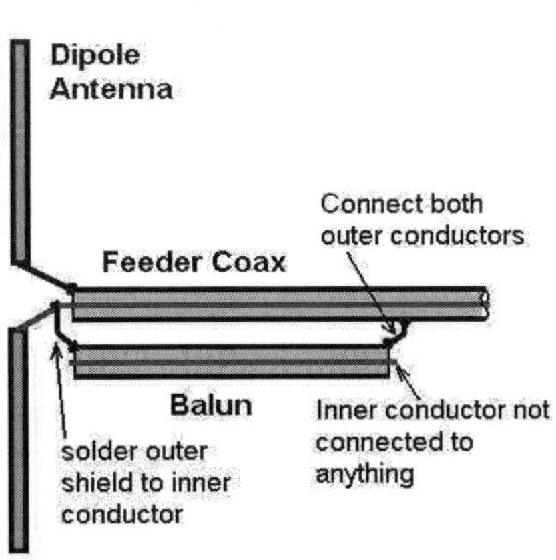

Making a Balun for a Half-Wave Dipole Antenna

Half-Wave dipole antennas are frequently used in airplanes, but they are balanced antennas and the coax feeder cable is unbalanced. It may be necessary to use a balun to convert the unbalanced feed to a balanced output. If this is not done, the coax itself may radiate, causing interference and distorting the radiation pattern of the antenna. Of course the antenna must have the right length for the frequency used to start with. The word "balun" comes from shortening "balanced-unbalanced". The 1/4 wave balun converts the unbalanced cable to a balanced output to drive the balanced dipole antenna. To make a balun, use the same type of coax as the main feeder coax. Only the outer shield of the balun is used, the inner conductor is not used for anything. To secure the balun, just strap it to the feeder coax with tape. To calculate the correct length of the balun, use this formula:

Balun length [ft] = 300 / frequency (MHz) * 0.68 * 0.25 * 3.28

For a frequency of 125 MHz, the balun length would be 1.338 ft = 300 / 125 * 0.68 * 0.25 * 3.28

Electrical System Redundancy

These days, many airplanes are loaded with equipment consuming electrical power. Have you thought about what happens if your electrical power source fails? If your flying is strictly VFR, it should not be an issue. However if you fly IFR and/or at night, an electrical system failure is very undesirable. Or how about that electronic ignition system? Will the engine continue to run? Basic redundancy is provided by having one alternator or generator and one battery. If either one of them fails, the other one will at least for some time continue to provide power. The problems you can have with such a system lie in the nature of the faults. If there is a short somewhere in the system, you can lose all electrical power regardless of how many power sources you have. If your electrical system is well designed, there should be no single point failure which can knock out your entire instrumentation.

The best backups are mechanical instruments independent of electrical power. GPS or primary flight displays with built-in backup battery provide an excellent means for redundancy. A handheld radio covers the communication side, but you may have to expect it to have a lot less range than your regular radio. If your alternator should fail, you need a means to alert you to this fact, otherwise the first hint you get that something is wrong is when your instruments are dropping off-line (dead battery). If you do not reduce the electrical load when you operate on battery only, the time you have power will be much shorter than if you shed loads. A low voltage light to alert you to an alternator failure is recommended. In a 14V system, as long as you have 13.5V to 14.5V, the alternator is working. If it drops below 12V your alternator just quit and the battery is draining. When you do your panel layout, think about what your essential equipment would be and how much power it will draw. If your load with only the essential equipment operating is 5 Amps, and you want to have an hour to get on the ground, your battery should be >5 Ah, preferable twice that because only a new battery has its rated capacity. If the battery is a few years old, the capacity can be much lower, half or 1/3 of what it is rated. The best bet is to replace the battery every two years. You can use a sealed motorcycle battery; they are fairly inexpensive and work well in airplanes.

A different way of providing redundancy is to use two busses. Normally they are not connected electrically. Each one is fed by a different power source and provides electricity to about half of your equipment. In the case of the failure of one bus, the faulty component (alternator) is turned off or the circuit breaker is pulled. Then a cross tie is closed to connect the two busses electrically. This type of electrical system typically has two alternators and two batteries, one for each bus. A simpler way would be to just have a second standby battery which is not normally connected to the main bus.

A flashlight is needed as backup for night flights. The problem is, you probably need it so rarely that when you do, the battery is dead. The best solution is to use a flashlight without a battery. LED lights are available with a hand crank. This is used to charge an internal capacitor or small battery. Some models can even be used to charge a cell phone battery, convenient if you are not in the vicinity of a power outlet on a trip. My favorite type is the dynamo pump light. You pump it a few times with one hand and it lights up for a long time.

Solar Panels

It helps keep the battery charged if you have a small solar panel in or on your airplane. Especially if the airplane is not flown frequently and is parked outside, it helps by keeping the battery topped off for better starting. A thin, flexible solar panel can easily be glued on top of the fuselage. It only requires one or two small holes for the wires. If you do not use a canopy cover, you can also place a solar panel on the glareshield. Connect it directly to the battery. Solder a diode into one of the wires to prevent the solar panel from discharging the battery if the sun is not up. The efficiency of flexible solar panels is about 5% - 9% at room temperature, but it drops the higher the temperature is. You can select the size you need by dividing the desired current in mA by 1.4 to 1.7 (mA/in^2). The resulting number is the size of the required panel in square inches. For example, if you would like to have a charging current of 100 mA, divide it by 1.6 mA/in^2 and find that the area you would need is 62.5 in^2. There are basically two types of solar panels. One type works best in direct sunlight (dark blue to almost black in color), the other type also works in diffuse daylight (grey in color). Select the type appropriate to your prevailing weather conditions.

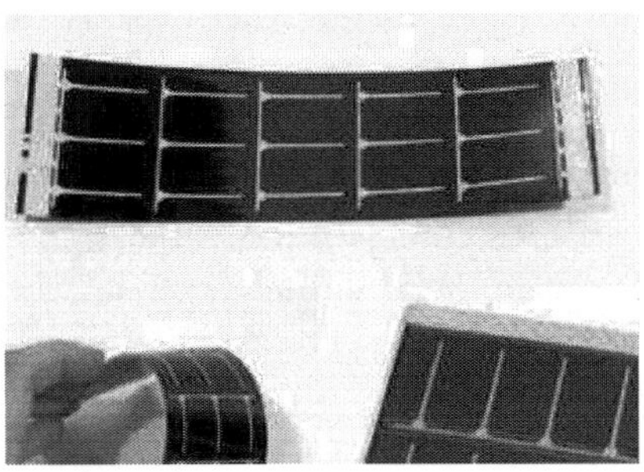

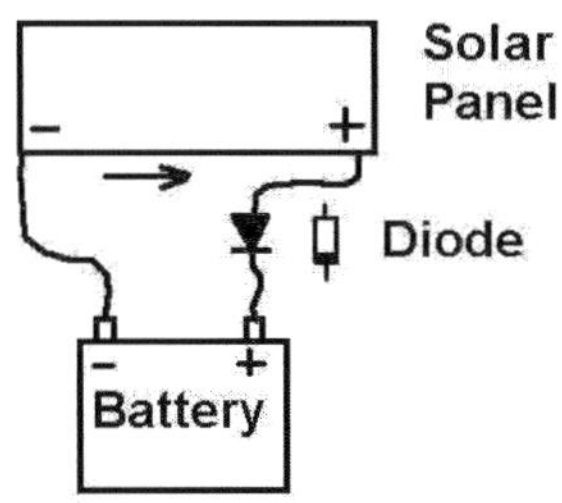

Radio Noise

When you turn your radio on for the first time and hear a lot of noise which has nothing to do with what the guy in the tower is trying to tell you, you may need a capacitor. Try to identify which system creates the noise. Prime suspects are electric motors. Turn things on and off until you find the culprit. Once you have identified the source, add a capacitor across the wires to the source.

Strobe Lights

Strobe lights can be an annoying source of noise in the headset. To avoid this, it is vital to ground all of the radio and audio components to exactly the same place as the strobes. The theory is that if they are grounded at different places, an electrical potential can develop between the grounding points, creating radio noise. The strobe case should be grounded to this same place with a dedicated 16 AWG wire. It is important to use a shielded, twisted pair of wires for the power leads, with only one end of the shield grounded. The shield is always only grounded on one end, for the same reason as before, to prevent any potential difference in the two ends of the shield, and thus a noise producing current flow. It is equally important to keep the power supply and the lights at least 6 ft away from any antennas or receivers.

Cutouts in Wingtips for Lights

If you want to reduce drag and mount you wing tip lights internally, you can modify your wing tips for it. It is easier if your wing tips are removable. Based on the size of lights to mount, mark the outline of the cutout on the composite wing tip. Keep in mind that to satisfy the regulations for the placement of position lights, they have to cover an angle of 110 .

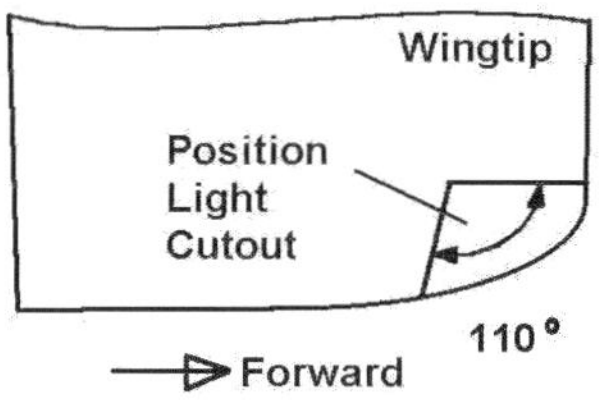

Cut along the marks with a dremel tool cutoff wheel. The piece that was cut out will be used to create the flanges for mounting the lens. If the thickness of the wingtip is different from the thickness of the lens, you need to address it now if you want the lens to be flush with the wing skin. If the lens is thicker, add material to the cutout piece along the edges to obtain the same thickness as the lens material. This can be made from several layers of tape, cardboard or anything else suitable. If the lens is thinner than the wingskin, consider using a thin gasket underneath the lens to get it flush. Once you have the required thickness, release the inside of the cutout piece with packing tape or clear tape and mount back in the place where it was cut from. The flanges will be layed up on the inside.

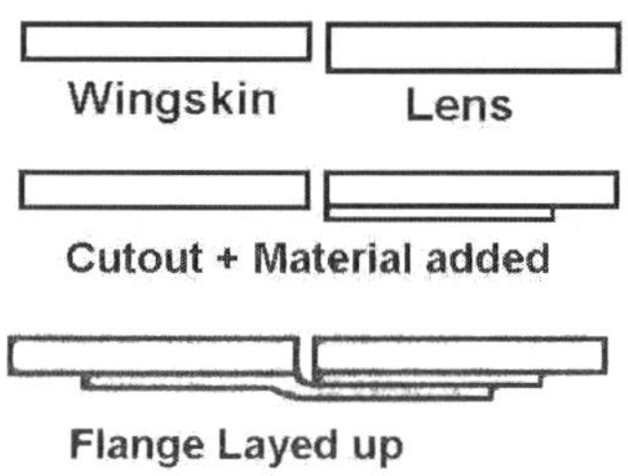

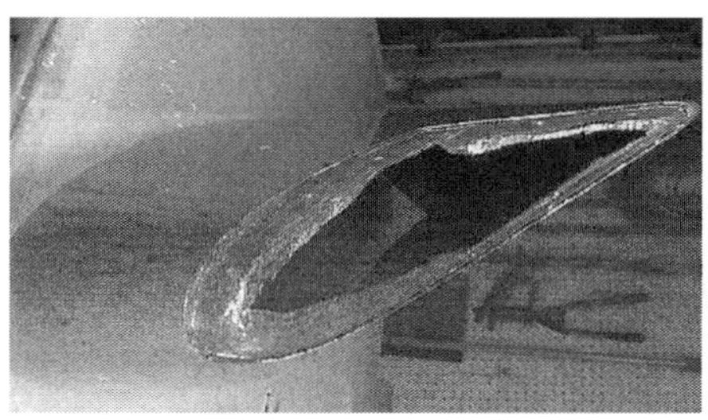

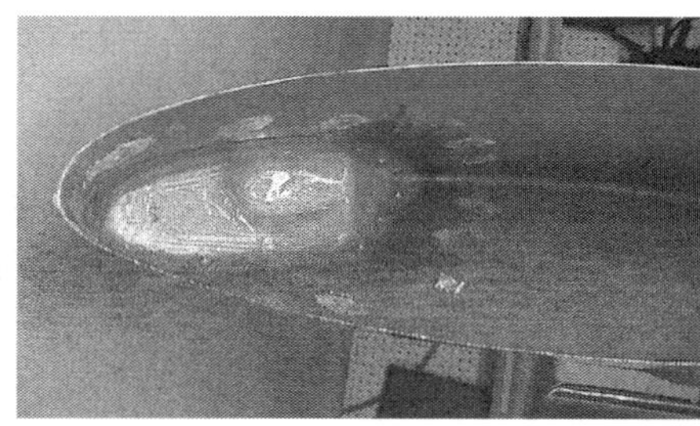

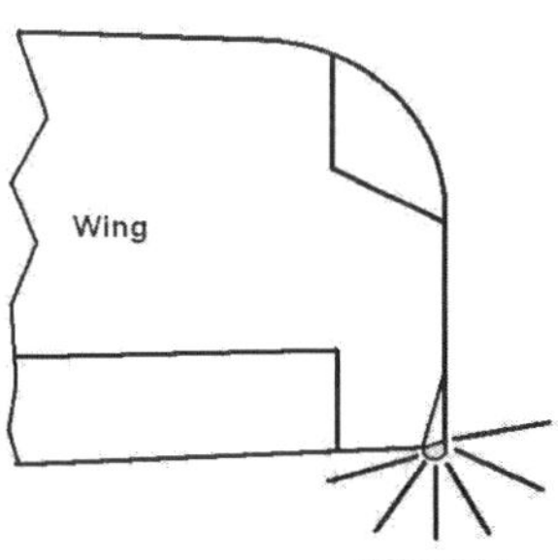

When the flanges are cured, as shown in the center picture, trim them to 3/8 inch to 1/2 inch. To create the inner mounting surface for the light, make a cardboard template cut to the shape of the wing tip. Clear-tape it in place and do a layup on the inside (right picture). Once this is cured, remove the cardboard, place the lens on the cutout, trim it to size and drill holes through the lens and flange for attachment. Install nutplates on the flange as needed to secure the lens. Attach pieces of shiny aluminum sheet to the inside for reflection and mount the position lights.

The rear position light (white), which is often installed in the rudder to be at the very tail end of the airplane, can be replaced by two rear-facing white lights on the wing tips. If you do not want to use the commercially available combination lights, which bolt to the wing tip and consist of a colored, forward facing NAV light, a strobe light and a white aft-facing position light, you can use two single white lights and install them on the aft tips of the wings.

Installations of Landing Lights

Landing Light Choice

Since there is no regulation about landing light location, quantity or intensity as there is for position lights, you are completely free to experiment. You have the choice between four different types of landing light types. The first, most widely used and least effective type is regular incandescent lights.

These are the same light bulbs you probably have throughout the house and in several places in your car. They have the advantage of being cheap, readily available in many different sizes and with different power consumptions and light outputs. The disadvantage is their poor light output at relatively high power consumption and short life. A lot of the power is simply converted to heat. All high power lights must dissipate a considerable amount of heat into the surrounding atmosphere. Thermal management becomes a major design issue when designing with any high power landing light. Their life is only a few hundred hours.

The second type is halogen light. They are related to incandescent lights because they have a glowing filament in common with them. Because they are filled with halogen gas rater than argon or nitrogen, filament evaporation, which is a problem of all incandescent lights, is reduced and allows operation at a higher temperature and efficiency. Their service life is similar to incandescent lights. They need to be handled with care. Do not touch the light bulb itself; any contamination will drastically shorten its life.

The third type, xenon or HID (High Intensity Discharge) lights, are based on metal halide arc-discharge technology. These have the highest light output and the lowest power consumption. I have tested them at night and found the light output of a single light high enough that I could have landed on an unlighted runway just using this light. Of course there are some drawbacks. They require a high voltage unit (igniter) and need a converter to produce it but the power consumption is lowest of the types of lights mentioned so far. The high voltage unit should be part of the bulb, eliminating high tension leads from the ballast as opposed to earlier designs. This means reduced EMI and thus minimum radio and avionics frequency noise. They also generate less heat than the other lights. They are more expensive than the two previous types ($500 to $550 for installation kits). They last from 2000 hours up to 3000 hours. The higher the output, the shorter will be the life.

The fourth type is LED lights, which are becoming bright enough to be usable as landing lights. They are fairly efficient and have a very long service life. There is not much practical experience with them as landing lights yet. The prices for one light (containing many LED's) vary from $500 to over $900, but its life should exceed that of the airplane even if you have it on all the time. The current draw for one 43W light is about 1.5A at 28V or double that at 14V. Because LEDs generate light only at a very specific wavelength and not over a broad spectrum like incandescent lights, they will not show up on an enhanced vision system because enhanced vision systems utilize infrared sensors. This should not be an immediate concern to most homebuilders.

Comparison of Efficacy by Power

Incandescent Lamp			**Halogen Lamp**		**LED Lamp**		**HID (Xenon) lamp**	
Power (W)	Output (lm)	Efficacy (lm/W)	Output (lm)	Efficacy (lm/W)	Output (lm)	Efficacy (lm/W)	Output (lm)	Efficacy (lm/W)
35	350	10.0					3200	91.4
50	700	14.0	1500	30	3500	70	5300	106
75	1200	16.0						
100	1700	17.0						

Luminous flux specifics the total light output in lumen. If you compare a light bulb with a garden hose, the luminous flux would be equivalent to the total amount of water spraying from the hose. Luminous intensity specifies the intensity of the light in candela. It is a measurement of the brightness you see when you look at the light from a certain distance. Using the garden hose comparison, this would indicate how much of the water hits a certain spot. This table shows a comparison of the light output in lumen (lm) for incandescent and HID lights and the change in efficiency for lights of different power ratings.

Installation Locations

Landing lights can be installed in the cowling, the wings or be attached to the landing gear on retractable gear airplanes. In all cases they should be easily accessible to replace bulbs if necessary. Some lights may require ventilation to reduce the heat buildup. A light centered with the airplane provides the illumination where it is needed the most: directly in front of the airplane. The only adjustment would be the angle of how far up and down it is pointed. For taxiing, it should light the taxiway about 50 yards ahead of the airplane in the normal ground attitude of the airplane. For landing, a different angle may be required depending on the angle of the airplane on final approach. If there is a large difference, you may need two differently adjusted lights for the two conditions. If the light is in the wing, it may also need to have a provision for lateral adjustment. If it is just pointed straight ahead, it will keep what is in front of you in the dark. Lights are available with different illumination angles. A taxi light should have a wide angle to light up much on both sides. A landing light can have a narrow, focused beam.

Installation of a light in the cowling means that the light is subjected to heat, vibration and potential leakage of fluids such as oil or coolant. This may shorten the life of the light. You will also need a connector for the wires to be able to remove the cowling. If you forget to reconnect them when installing the cowling, you may find out only much later, when you need it the most. One advantage of this location is that the light is not in direct view of the pilot like a wing light and therefore offers reduced glare. Access to the light in the cowling is of course very good and high heat output by the light and ventilation should not be issues.

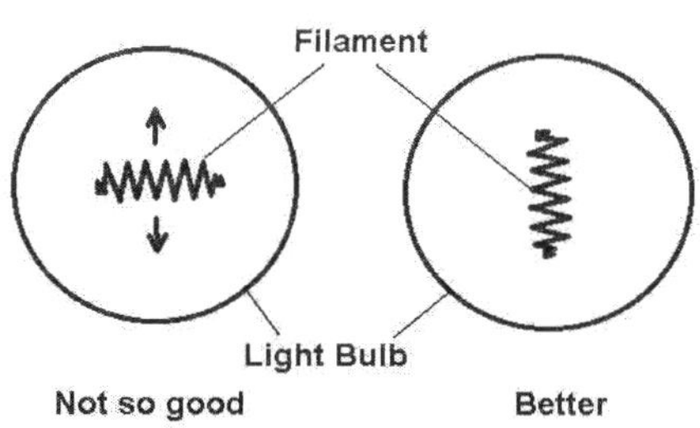

To make a cowling landing light installation, start by determining the location. You need to have enough space for the light assembly and at least 1/2 inch clearance to the engine. The best location may be slightly off center. You will need a piece of clear plastic as the lens cover. It is recommended to bond this to the cowling with high temperature silicone. Screws will usually lead to cracks in the lens sooner rather than later because cowlings are rather flexible while the lens is not. This means that bulb replacement will have to be done from the inside of the cowling. Cut the lens to the shape you need. Lay it on the cowling and trace its outline. Cut a hole in the cowling along the outline but slightly inside the line with a Dremel cutoff wheel. Sand the edges until the lens fits into the hole. Cover the lens with tape to protect it and tape it in place in the hole, flush with the outside, for the layup of a joggle. Make sure it does not move, otherwise the joggle depth will be incorrect. Surface prep the inside of the cowling and layup a flange of the width you want (about 1/2 inch). Use the same number of plies as the cowling is made from. This will also provide a reinforcement of the hole. Once the layup is cured, remove the lens and trim the flange to its final width. Now you can build the light attachment structure. Route the wires from the light along the side of the cowling and layup two plies over them to hold them in place. Once you are done, coat the inside of the cowling with protective paint.

If you are using incandescent light bulbs, it is best to orient them with the filaments vertically rather than horizontally. Loads and vibration will cause the horizontal filament to break sooner than the vertical ones.

If you are placing the landing light in the wings, make sure there are no structural concerns in the location you want to cut a hole into. Typically the further outboard, the less effect it will have on the structure, but this moves the lights further off center. It is best to ask the designer about the suitability of the planned location. Another consideration is the shape of the lens cover. If you have a constant chord wing, it has the same shape everywhere. If the wing planform is tapered, different locations require different lens cover shapes. If you already have a lens cover, it will most likely only fit well in one location. In the wing, access to the light is by removal of the lens cover, so it has to be attached with screws. If your airplane is composite, you need to be concerned with heat buildup. You need to space the lights away from the composite structure with aluminum brackets and allow for plenty of openings to allow airflow around the lights.

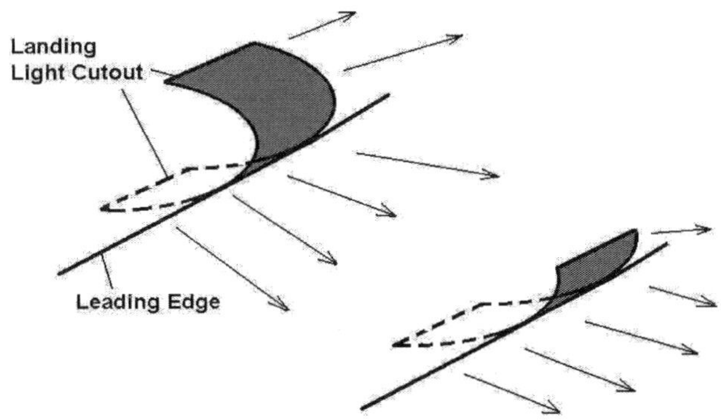

Before you make a cutout in the wing leading edge, make a template of the size you are considering from paper and tape it in the location you are planning to use. Sit in the cockpit and look at the template. How much of it can you see? Remember that this area will be lighted up and if you look at it in the dark, it will blind you. If necessary, move the cutout towards the bottom of the wing to reduce the amount of light you can see (the shaded area in the sketch). There is not much reason to light up the sky above the airplane. The shaded area in the sketch is the amount of lighted-up lens you would see.

The picture below shows the cutout made for one installation. There are four mounting holes with nutplates bonded behind them. Because of the depth of the light assembly, another hole was cut in the web behind it. This also allows more airflow for cooling. The right picture shows the parts including the attachment bracket for this installation. Grommets are used to space the rear aluminum mounting plate out from the attachment to the web. The lights are spaced out from the rear mounting plate by two L-shaped aluminum brackets.

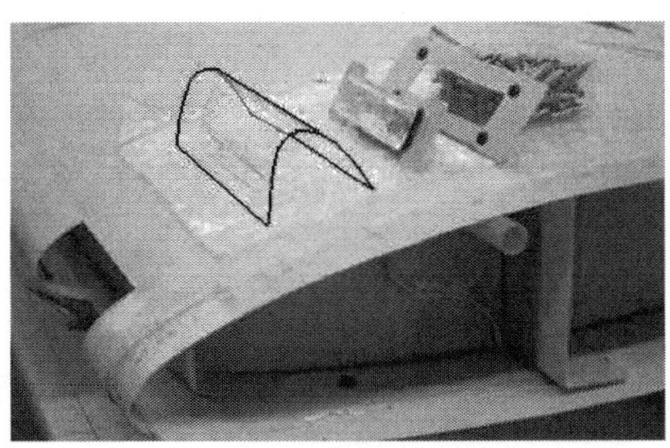

This provides a longer path for any heat to be conducted from the lights to the composite, and should limit the temperature that the composite material sees to something reasonable (<150ºF). Most of the heat though will be radiation heat. It helps to paint the structure surrounding the lights with high temperature, reflective paint.

Fabricating the Flange

Because there only is access through the cutout and no access from the rear of a closed wing, the lens mounting flange needs to built in a different manner as previously described access panel flanges. It is recommended to use sheet wax to set the depth of the flanges equal to the thickness of the plexiglass cover. Sheet wax is a product that comes as a sheet of wax (1/16 inch, 1/8 inch thickness, etc.) with adhesive on one side. It is available from McMaster Carr, for example. If you buy the 1/16 inch sheet, you can use two layers to get the 1/8 inch needed one application and a single layer for another. Cut 1 inch wide strips of the wax. Using the rolling "pizza cutter" works well. Cut 3 inches wide strips of thin flexible aluminum (flashing material from the hardware store works well). Tape the aluminum around the opening with a uniform 1 inch overhang (see picture). Peel the protective paper off of the wax strips and carefully apply a layer to the inside of the aluminum, where the aluminum overhangs the fiberglass opening. Repeat if using 1/16 inch wax and 1/8 inch Plexiglass. This is the tooling for the layup. Make sure the aluminum strips are firmly in place and do not sag. For the layup, prepare 2 inch wide, 4-ply pre-wet strips, and apply them over the wax and onto the surrounding fiberglass on the inside of the wing structure. This is done by reaching through the opening.

After curing, remove the aluminum and the wax, clean up the surfaces, and trim the overhang to your desired value. 3/4 inch is a good final value, but you may cut it down to as low as 1/2 inch on the ends. Attach nutplates to the flange to hold the lens in place.

Test the lights on the ground by turning them on and monitor the temperature of the structure. If it does not get too hot in 30 minutes, you should be fine because normally you would have airflow over the wing which helps cool it. If you find the temperature gets too hot (150 F / 65 C), you may need to add holes or slots to the lens at the bottom and allow the hot air to vent somewhere near the top.

Other Systems

Assistance With Systems Design

If the instructions in your builder's manual are a bit vague on systems installation or you want to make your own modifications, get hold of parts or service manuals of a certified airplane. This can show you how it has been done and what works. Piper or Cessna service and parts manuals can relieve a lot of the guess work regarding many systems installation questions. Try to imitate certified aircraft systems when possible.

Ergonomic Rudder Pedal Design

The design of most basic pedals features an upright "T" pivoted near the floor and moving in a vertical arc (left and center picture). Being more or less vertical means the pivot of the pedal and the point your foot pivots on (i.e. your heel where it rests on the floor) are some distance apart and therefore your foot either slides a little on the pedal or your heels slide along the floor when applying rudder (left picture). To improve this, move the pivot of the pedal closer to where the pivot of your foot is. Put in an extension tube running aft from the point where a pedal would usually be pivoted, under the floorboards to nearly the same point your heel would be (right picture). The feel of the pedal will be a lot better; the foot stays in contact with both the floor and the top of the pedal, with no sliding motion anywhere.

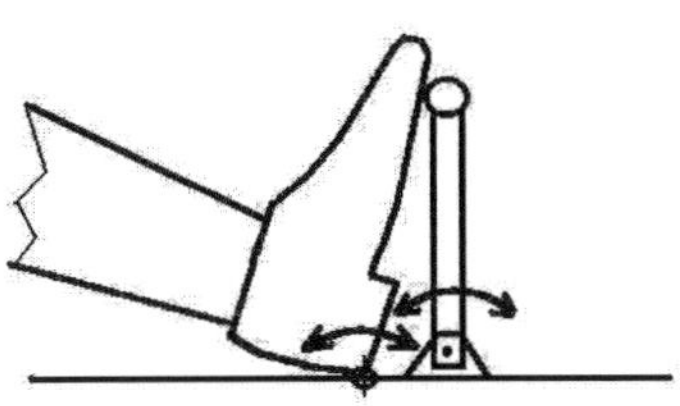

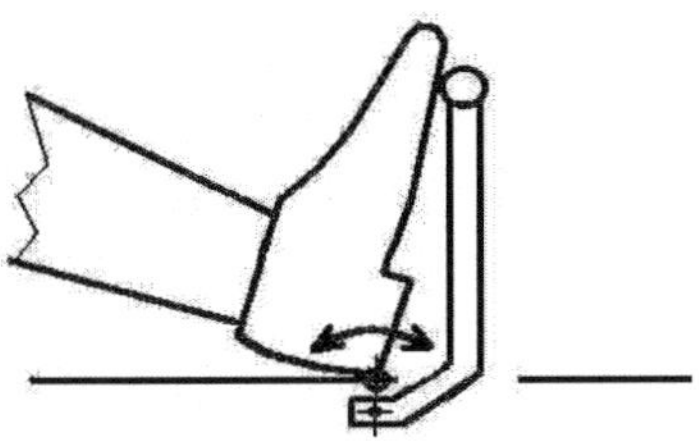

Brakes Tubing Installation

When you install hydraulic brake lines, try to install them with a continuous slope towards the brake fluid reservoir, which should be mounted up high. This helps avoid air bubbles when filling the lines and will make bleeding the brakes a lot easier. To bleed the brakes, open the fitting at the body of the brake and slide a rubber hose over it. Attach the hose to a squirt can filled with brake fluid to slowly fill the lines from bottom to top. The air in the lines will be pushed out and is replaced with fluid. The most common brake fluid is hydraulic Fluid MIL-H-5606. Aluminum brake lines are lighter than hose assemblies. If the method of using aluminum aircraft tubing conforms to the Standard Aircraft Handbook, they generally have a longer service life than plastic tubing, but they are also more susceptible to vibration damage. Plastic tubing (for example Nylaflow tubing, rated to 1000 psi) is suitable in many cases as brake lines. Make sure the working pressure for the tubing you select is at least 300 psi. Plastic tubing is not as durable as metal tubing, but if there is much vibration or relative motion, the flexible plastic tubing is the best compromise between cost, weight and durability. The fittings used for your installation depend on the material and size of the tubing. Use standard aircraft flared fittings for metal tubing or brass Poly-Flo fittings for Nylaflow tubing.

Brake Actuation

This is a collection of examples how builder have installed their brake actuation at the rudder pedals. Most prefer to use toe brakes. You can better apply controlled force with the tip of your foot than with the heel while sitting down. If your rudder pedals are a welded steel tube assembly, the brakes can be actuated by a bracket cut from an aluminum L-profile rotating on the tube as shown in the left picture. The vertical brake cylinder is mounted rigidly to the floor near the axis of the rudder pedals. The floor in this location may need to be stiffened to react the loads from the heaviest braking without too much flexing. The center picture shows a rudder pedal with an integral arm for the brake cylinder, which is horizontal in this case. The rudder pedal pivots on the bottom hinge for brake actuation while the pedals are mounted on a horizontal bar which pivots on a vertical axis between the two pedals for rudder actuation. This arrangement minimizes the space required vertically in the cockpit.

The right picture shows the arrangement in a canard airplane with rudders in the winglets. Here one rudder is actuated at a time or both simultaneously as speed brakes. The brakes are actuated if the rudder pedal is pushed beyond a certain point. A bolt is inserted thru the actuator and thru a hole in the rudder pedal. As the pedal is compressed, the bolt slides in a slot until it reaches the end of the slot. Once the bolt hits the end of the slot, the master cylinder is actuated by further pushing on the rudder pedal. The slot is designed to allow the rudders to be fully deflected at the end of the slot. The adjustable foot pegs can be seen in this view as well. The square tube slides through its guide and can be bolted in various positions.

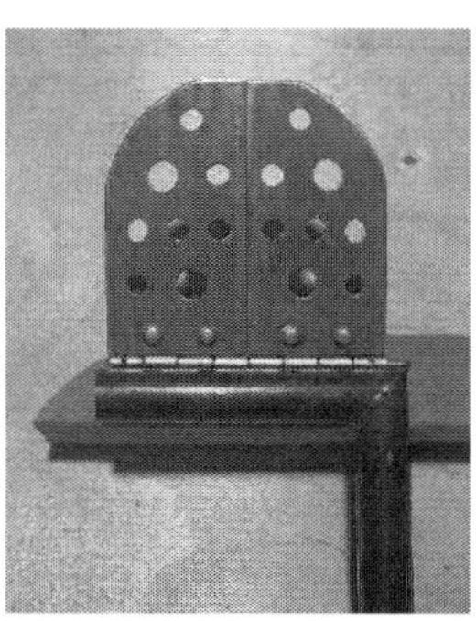

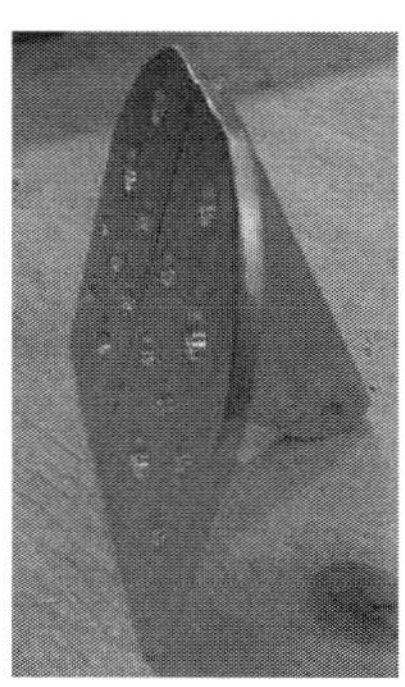

In this example the brake pedal, which is cut from an extruded aluminum T profile, is attached with a piece of piano hinge to the top of the rudder pedal.

When you design your rudder pedals, it is a good idea to add a fence on both sides which keeps your foot from sliding off the pedal sideways. Otherwise it may happen, especially when you are busy, that you may be pushing on something other than the rudder pedal or even the wrong pedal.

Adjustable Rudder Pedals

One method of adjusting the rudder pedal distance to the seat was shown in the previous paragraph. A common method used in many gliders is the sliding assembly. The rudder cable is routed through S-shaped tubes as shown in this picture. The rudder cable is attached somewhere forward of the pedal to a fixed structural point. The bottom part of the S-shaped tube ends in the pivot point of the pedal. When the pedal is pushed forward, the top part pulls on the rudder cable while the bottom part of the cable does not move. To adjust the rudder pedal assembly, the cable is free to slide through the S-shaped tube while the pedal assembly is moving forward or aft.

A simpler method with a more limited adjustment range just changes the cable length while keeping the pivot point in the same location. This changes the angle of the rudder pedal. The design shown in this picture allows quick adjustment by sliding the bolt into a different slot.

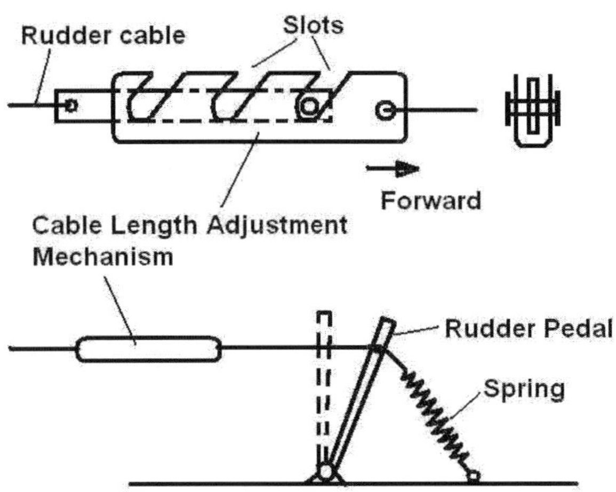

Wire and Cable Conduits

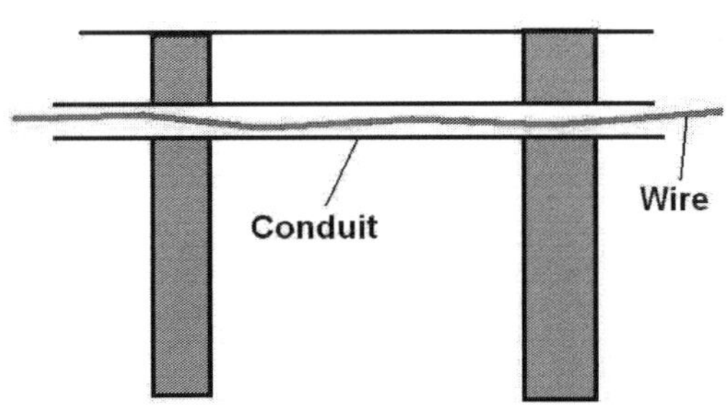

There are often places where you need to run wires or cables in an area which after the airplane is completed may be inaccessible. A replacement of the installed wires in case of a problem could become very difficult. Identify such area early and use a conduit. The purpose of the conduit is to allow wires to be pushed through from the accessible end without opening up the structure. The size of the conduit of course depends on what you want to insert, but it should not be much larger than necessary to prevent the wires from kinking in the tube. A nylon tube usually is suitable as a conduit. A corrugated tube or hose would be less suitable. Another suggestion applies to wings where the pitot tube is in the wing. Usually after the wing is closed the tubing is not accessible anymore. A leak in the system would be hard to repair. Just in case, you could run a spare line next to the first one which can be used if there is a problem with the first one.

Fuel Pickup in a Tank

These pictures show two methods of building the fuel pickup inside a tank. The fitting in the right picture is mounted on an access panel and is held by a bracket to keep it from rotating because once the tank is closed, it is accessible only from the outside. The tube attached to the fitting is bent to move the fuel pickup near the bottom of the tank. Slots were cut into the tube at the end to prevent larger particles from entering the fuel line. Make sure the size of the slots is appropriate for your fuel flow requirements. If they are too small, the engine will not get enough fuel to run properly.

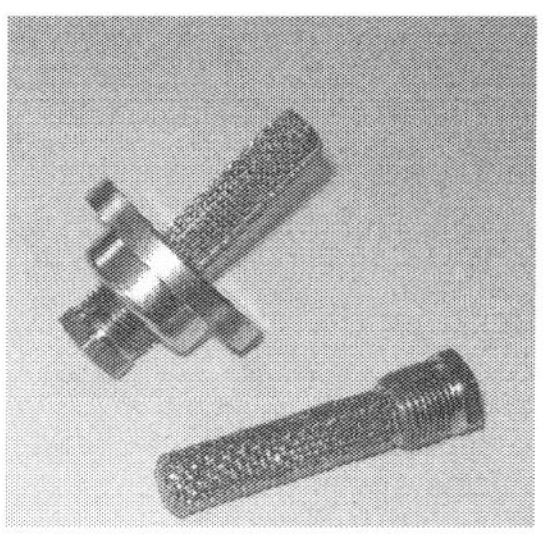

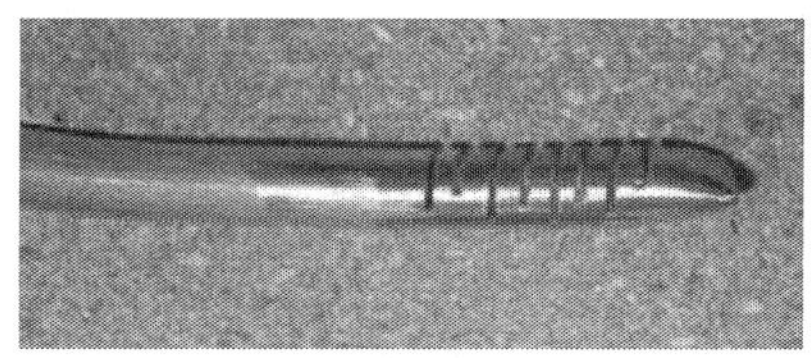
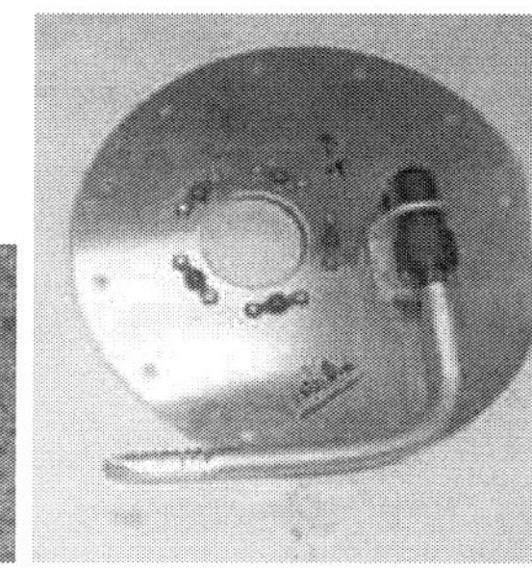

A different style pickup is shown in the left picture. It consists of a flange with a pipe thread and a removable screen. The flange will be welded or bonded to the tank. This has the advantage that the screen is removable for cleaning without having to open the tank.

Fuel Tank Material

Typically you have three choices: welded aluminum tanks, bladder tanks or integral tanks. The third choice is further divided into two different materials based on your airplane's structure: metal or composite. Integral tanks in metal airplanes rely heavily on sealant for leak-proofing. This is a messy job because the fuel tank sealant, a polysulfide two-component material, is about as nice to work with as liquid chewing gum. The sealant may not stick well to smooth aluminum surfaces. Scuff them with scotchbrite to promote the adhesion. Once the tank is sealed, it should stay sealed for many years. Unfortunately the sealant deteriorates after many years, depending on the conditions. High temperature or alcohol in the fuel may accelerate the deterioration. Then the job of scraping off the old sealant and reapplying new gunk is even worse, even if you have access holes in the tank. Integral composite tanks are easier to seal if you are willing to work very carefully while building them. Composite material is slightly porous by nature. When building the tank you have to use excess resin and a very thin ply as the inside ply. If you did a good job, a composite tank can be just as leak-proof as a metal tank. Epoxy and vinylester resin have good resistance against fuel, but it is advisable to coat the inside of the tank with sealant to protect the structure from the softening effects of the fuel, especially if there is alcohol in the fuel. Once it is properly sealed, it will usually stay sealed and should cause no problems. If the tank walls were improperly sealed, the material's porosity will continue to cause problems and leaks. The only fix is to clean and sand the inside of the fuel tank, layup another ply with excess resin and reapply sealant. The best way to avoid this trouble is to use welded aluminum tanks instead. This adds weight over an integral tank, but chances are good that they will never develop a leak regardless of which fuel is used. The nice thing is they can be leak-checked before installation. Bladders are rarely used on homebuilts, because they have to be custom made to fit the particular application, an expense usually not worth it. Unless you can find a bladder which happens to fit your plane and you are willing to accept the extra weight, integral tanks are the better choice.

Grounded Fuel Filler Ring

When the fuel truck drives up and the refueller connects the ground wire to the exhaust stack of your engine, will it do any good? Is your fuel filler ring grounded to the rest of the airplane? If your airplane is built from metal, the answer is probably yes. If your airplane is made from composite or wood, the most likely answer is no. If you have not grounded the fuel filler ring, static electricity can build up during the fueling process and a spark may jump from the airplane to the fuel nozzle and set the fuel fumes on fire. To prevent this, a path must be provided to allow the static electricity from the fuel tank area to bleed off to the fuel truck potential without sparks.

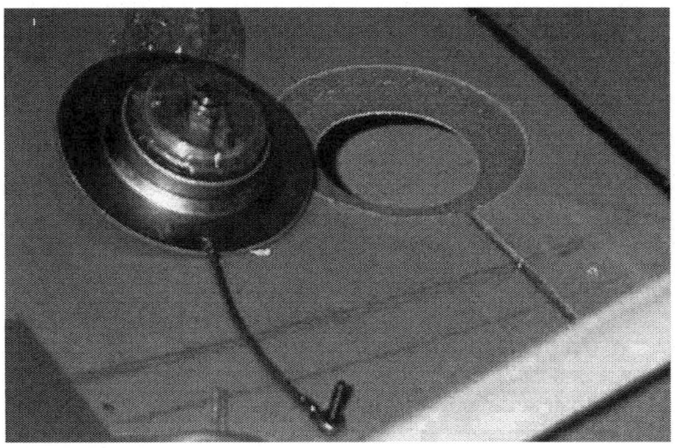

A ground wire needs to be attached to the fuel filler ring and routed to a location outside the tank where a connection to aircraft ground can be made. While you still have access to the inside of the tank during the wing build process, cut a hole the same size as the fuel filler ring through the whole skin and concentric with this a hole for the flange of the filler cap in the inner face sheet at the location where you want the fuel filler cap. Remove the foam, leaving a flange of fiberglass for the filler ring to sit on. By doing this work from the inside, there will be less sanding of skin upon the completion of the wing. Drill a hole through the flange. Prepare a ground wire from a piece of braid with spade lugs crimped on both ends. It should be long enough to reach from the flange to outside the tank area. Rivet the braid onto the rim of the flange. On the other end attach it to a small screw. Cut a slot for the braid to fit into the inner skin. In the picture you can see the lines marking the fuel tank end rib. It is close to the filler cap so the wire can be fairly short. Place the assembly into the hole and the braid into the slot, cover it with micro to make a smooth transition, and add the appropriate number of plies from the inner skin to repair it. The screw will be sticking out and a ground wire can now be connected to it.

Fuel Cap Retainer

A simple way to keep a fuel cap from getting lost if it is left open by accident is adding a retainer. The retainer is a triangular piece of aluminum which is attached by a short chain to the fuel cap. Both sides of the triangle are longer than the diameter of the opening. The retainer hangs on its chain inside the fuel tank when the fuel cap is in place. It allows the cap to be removed, but prevents it from dropping and disappearing if the refueler is careless. The fuel cap can be completely removed by turning the retainer sideways to get it through the opening. Existing fuel caps can be retrofitted with a retainer by drilling a small hole somewhere in the bottom of the fuel cap and securing the chain with safety wire.

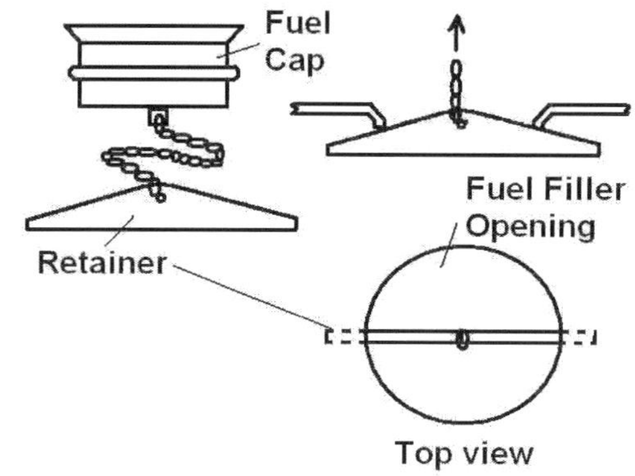

Fuel Tank Vents

Every fuel tank, even a header tank, needs to be vented to allow air to flow into the tank replacing the fuel consumed by the engine. Without a proper vent (tank sealed), the fuel pump drawing the fuel from the tank will create a lower and lower pressure in the tank. The fuel pump, once it is unable to overcome this low pressure, pumps no more and the engine quits or the tank collapses. For that reason, it is preferable to have the vent opening in a high pressure rather than a low pressure area on the airframe. Remember the pressure on top of the wing is low, underneath the wing it is high. In most other areas of the airplane, the pressure differences from static are minor from the fuel vent point of view. The shorter a vent line, the better and the smaller can be the diameter, although the smallest size should have at least 1/4 inch (6 mm) inner diameter. The vent line should exit the airframe in a manner that fuel leaking from it easily clears the airframe.

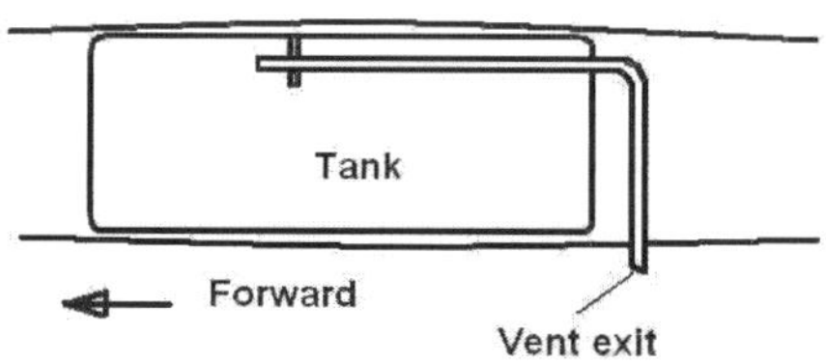

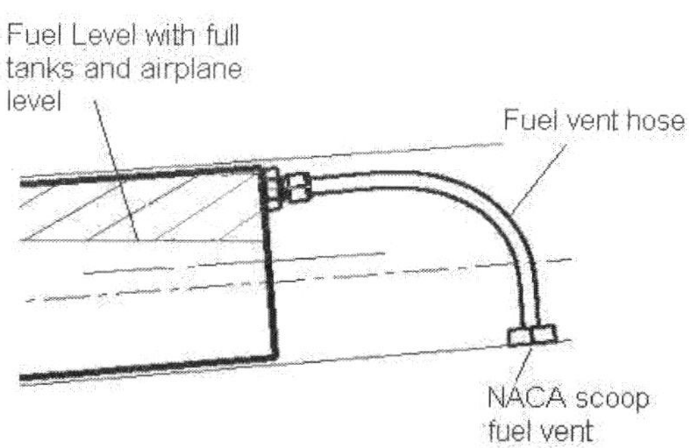

This very simple vent line drawn in the cross section of the wing (left picture) is a piece of tubing inside the tank. One end is near the highest spot in the tank, the other end exits at the bottom of the wing and protrudes about 1/4 inches past the skin. It is chamfered towards the front to provide some ram air pressure. This vent arrangement is best if the wings are removable and are rotated leading edge down for storage because fuel cannot drain from the tank vent in that attitude. Rather than exiting the tank on the aft side, the vent line can also be attached to a fitting in the outboard rib with a line to the wing tip (right picture). This has the advantage that the exit is higher than the tank on airplanes with dihedral. This helps prevent the loss of fuel if the tank is full and the airplane is parked on a sloped ramp. There should not be any dips in the vent line because trapped fuel in a dip can restrict the function of the vent. A little more work is a NACA style inlet for a fuel vent, but it is less likely to become clogged by ice or insects than a ram-air exit. The one shown here has a piece of flared tubing bonded in to which the hose is attached.

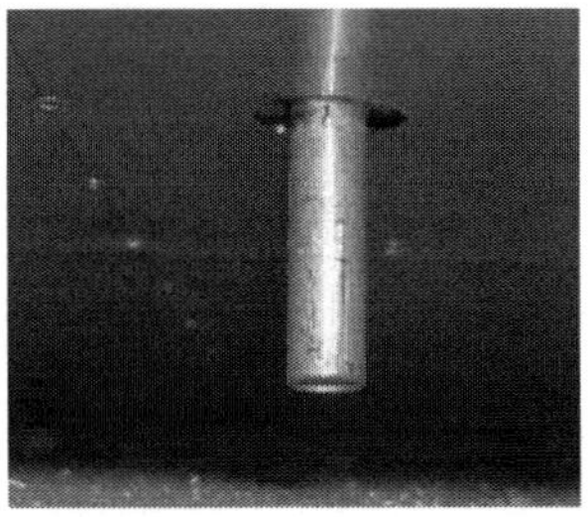
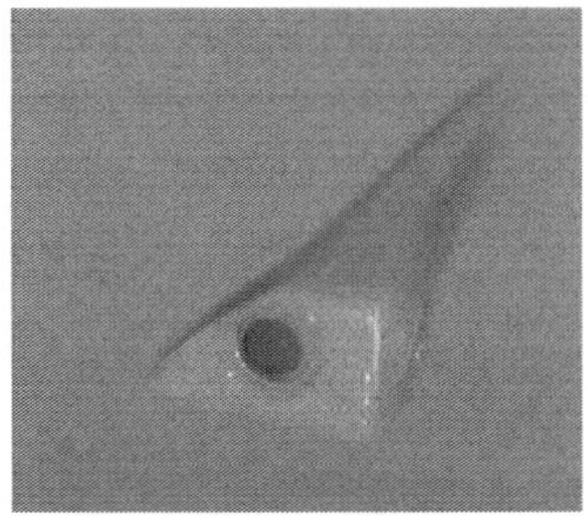
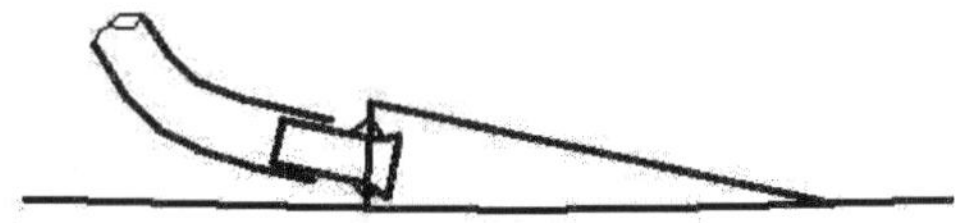

If your fuel vent exit looks like in the left picture, add a rubber grommet around the tube at the skin penetration.

If you are considering adding a check valve at the wing tip end of the vent lines to avoid losing fuel on a sloped ramp, think about a few things. A check valve will have to be open to let air flow into the tank as the fuel is consumed. It must not be a major restriction to the air flow. It may work, but remember that the vent system must also let air out of the tank. If the airplane is parked on a hot ramp and the fuel expands, the fuel vapor can build up to a lot of pressure if the tank was sealed by the check valve. This could easily rupture the tank. There are valves that will allow air to pass while blocking fluids or a small bypass orifice could be added to limit the pressure buildup. Adding a valve is of course more complicated and more expensive. Remember KISS and Murphy's Law. Check valves add complexity and failure modes.

Slosh Box Shape

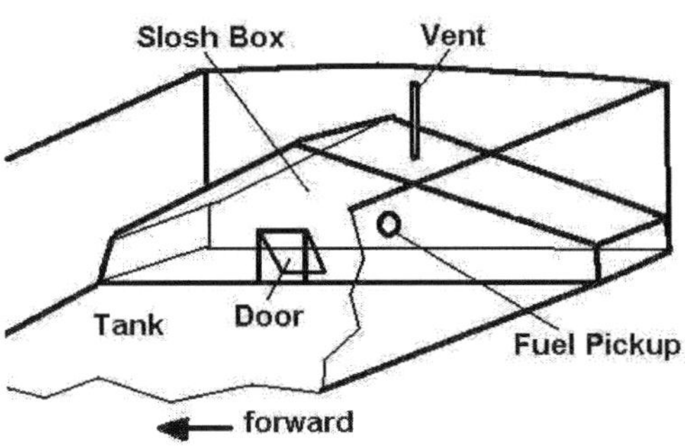

A slosh box in a wing tank is an enclosed space around the fuel pickup which holds about a gallon of fuel. It is integral with the fuel tank and prevents the fuel from unporting if the fuel level in the tank is low and turbulence is sloshing the fuel around. It should keep most of the fuel near the pickup whether the airplane is in a steep climb, descent or sideslip attitude that would otherwise cause the fuel to temporarily flow away from the pickup. A flapper valve at the outboard side of the slosh box swings open to the inside to let fuel flow in, but closes when the fuel tries to flow out. The most effective shape has proven to be sort of roof shaped. This keeps the fuel at the pickup in most unusual attitudes. A small vent tube allows air to escape from the box when fuel is flowing in.

Fuel Gauge Calibration

Regardless of the type of fuel gauges you have, you should calibrate them. Depending on the design of your airplane, its attitude on the ground and the attitude in flight and therefore the calibration may be quite different. If you want the fuel gauge to read correctly in flight, level it accordingly for the calibration. Start with the tanks empty and fill them **slowly** while recording the reading on the gauge. If your tank has baffles or ribs with small holes, it is very important to wait some time each time you add fuel to allow it to distribute. If you fill it too fast, the fuel may not have reached the sender when you read the gauge. The result would be a gauge that is reading high, something that could really hurt you. The increment in which you fill the tank depends on its size, but should be not more than ¼ of the tank volume at a time. If you have capacitance type fuel senders, calibrate them with the fuel you use the most. Autogas results in about a 25% higher reading than Avgas!

Wheel Pants

Wheel pants can reduce the drag and increase the speed of a fixed gear airplane considerably. Of course it depends on the overall drag of the airplane what the percentage of the drag of the wheels is. With a fairly clean airframe, like an RV8, the increase in speed with wheel pants may be 10-15 kts. With a higher drag airplane like for example a Piper Cub, you might only gain a few knots.

There are generally two methods how wheel pants are attached. The simplest wheel pant is a single piece, which fits over the wheel and attaches to the gear leg with a bracket. This picture shows how this leaves the brake and some of the gear leg exposed. The other picture shows a different brake orientation and wheel pant attachment bracket on a Pulsar gear. On the opposite side of the wheel, there typically is one screw in a hole tapped into the axle, supporting the wheel pant.

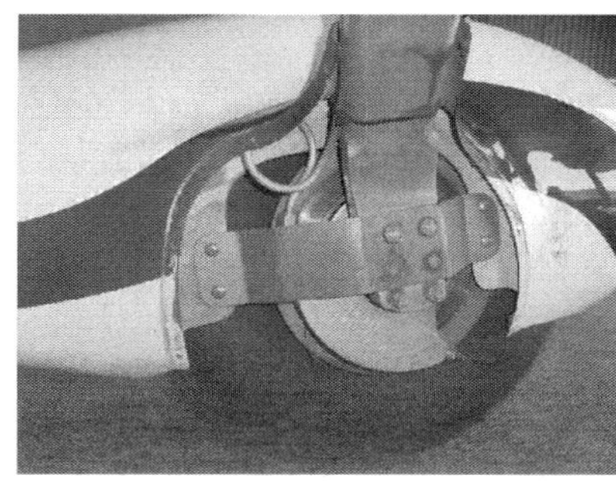

To lower the drag even further, the brake can be enclosed as well, but keep in mind that an enclosed brake gets hot quicker and stays hot longer than one exposed to some airflow. One builder has created a separate fairing which is attached to the existing wheel pant and covers the brake assembly and has a nice transition to the gear leg.

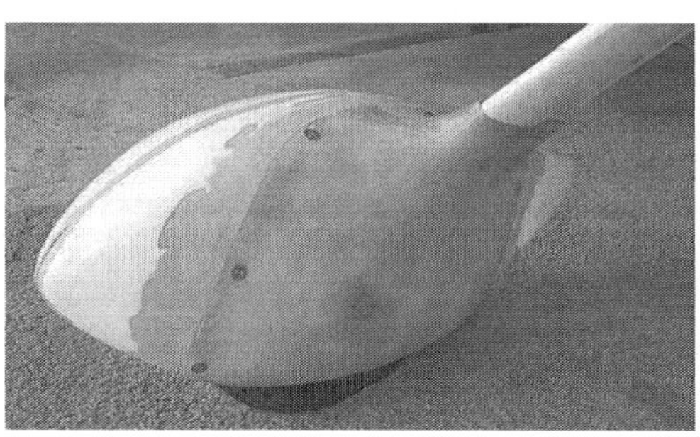

The second picture shows work in progress modifying a one-piece wheel pant into a two-piece wheel pant enclosing the brake. The support bracket is made from aluminum and is attached to the gear leg by the bolts holding the axle to the gear leg. It has four nutplates installed.

The other style wheel pant consists of a front and back half, which are joined at the thickest part with a number of screws.

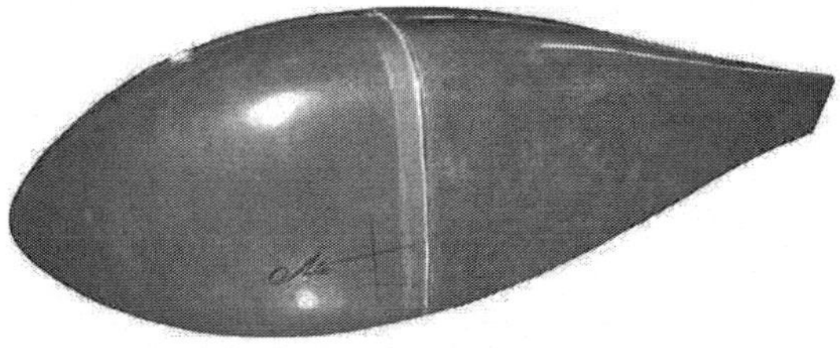

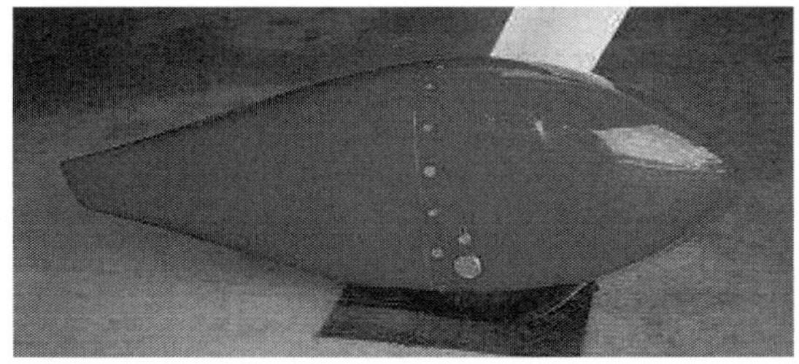

Wheel pants made from composite can be built fairly light. One ply of 6 oz/yd^2 carbon is actually sufficient for a small wheel pant, but if it is that thin, it easily deforms if something bumps into it which will cause cracks. Two plies are more durable. At the attach points and where countersunk screws are used, build up more plies, at least 4-6 plies. The fiber orientation is not important, except that the plies should have the fibers going in different directions.

The wheel pant should be large enough to leave the tire some space for deformation. Tires, made from rubber, are very flexible and can deform considerably during a firm landing. You should leave about 3/4 inch space between the top of the wheel pant and the top of the tire to avoid rub marks or a hole in that spot in the wheel pant. The hole at the bottom through which the tire protrudes must be slightly larger than the tire diameter and width if you have the one-piece wheel pant style, because it is installed over the top of the tire. If you have the style which splits in front and back half, the hole can be smaller, but you should still leave about 1/2 inch clearance minimum around the wheel. To align the wheel pant, prepare strips of wood or foam of the width of the clearance you want and tape them to the top and the sides of the wheel. These will be the guides to ensure that you have the minimum clearance. Look from the back and sides to make sure the pants are symmetrical with each other. Draw the centerline of the airplane on the floor and measure the distance to each main wheel. This distance may not be exactly the same if the gear is not deflected exactly the same. If there is a difference in distance from the centerline to the left wheel to the distance to the right wheel, use this as a correction when you measure to the leading edge of each wheel pant and the trailing edge. It will help you assure that the wheel pants are not toe'd in or out.

Tires pick up all kinds of debris and sling it around. For that reason there should not be too much open space in a wheel pant where the debris can collect and add weight to the airplane. Add a bulkhead behind the tire to close off that space.

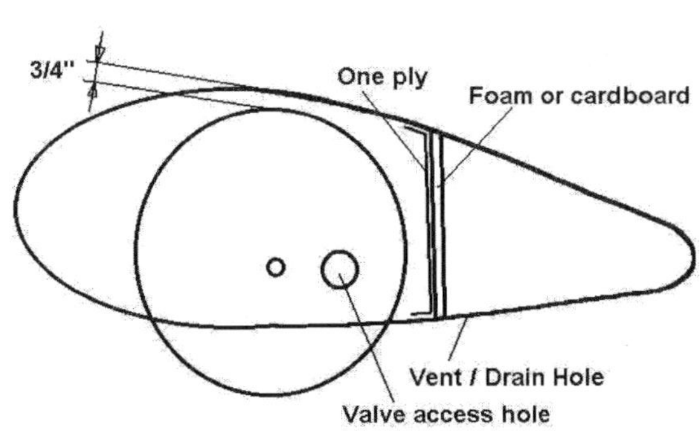

It is best to have a hole for reaching the valve stem for tire inflation rather than having to remove the wheel pant every time. This hole can be closed with a plug which snaps on and is easily removable (right picture). The alignment of the wheel pants depends a bit on the airplane. Of course they should be aligned as much as possible with the airflow, but if your airplane is a tail dragger, the rear part of the wheel pants must also clear the ground with the airplane sitting in its normal ground attitude.

Low Drag Tie Downs

There are generally two methods to reduce the drag of tie-down rings. You can make them removable or retractable. Removable tie-downs are very simple. They consist of a ring which is threaded into an inset or nutplate in the wing or tail. It is probably not necessary to further describe how to do this. If the designer of your airplane has not specified the location for the tie-down hard points, select a place close to the spar, next to a rib in order to have structure which can handle the loads. **Just do not drill a hole in the spar cap!** That would substantially weaken it. In the fuselage, the best location would be next to a bulkhead, but it may require additional structure to reinforce it (a doubler or extra plies for example) if there was no provision for it.

The loads which a tie down point may see depends on the size, the weight of the airplane and the attitude in which the airplane sits on the ground. The following table is for a nose gear airplane. The wings of a tail dragger parked into the wind create more lift so the vertical wing tie-down loads may be as much as 50% of the weight. You can use the following table to estimate the force the structure of the tie-down points may have to support. It is divided into vertical (straight down) and lateral loads for wing and tail tie-downs. The loads are calculated as the percentage of the empty weight of the airplane. To give you an example, the loads are also calculated for an airplane weighing 1200 lb.

	Vertical		Lateral	
Wing	25%	300 lb	12%	144 lb
Tail	15 %	180 lb	15%	180 lb

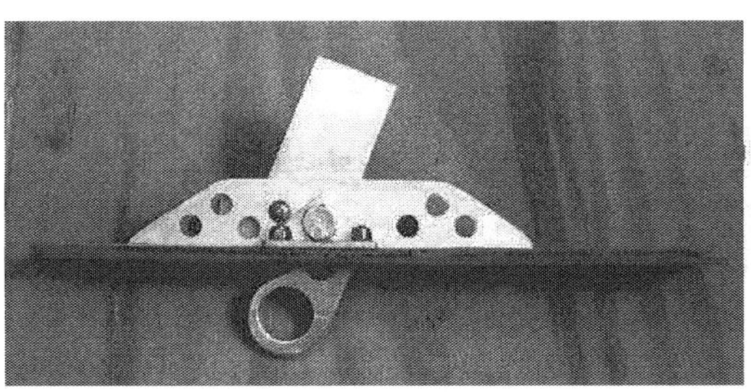

A retractable tie-down is a bit more work and may add more weight and require more space. The one described here is self-retracting, based on the principle that a counterweight balances the ring to retract on its own. The same effect could be achieved by adding a spring. Since most tie-downs are in inaccessible locations after the wing or fuselage is closed, the counterweight should be the preferred option because a spring is more likely to break and difficult to replace.

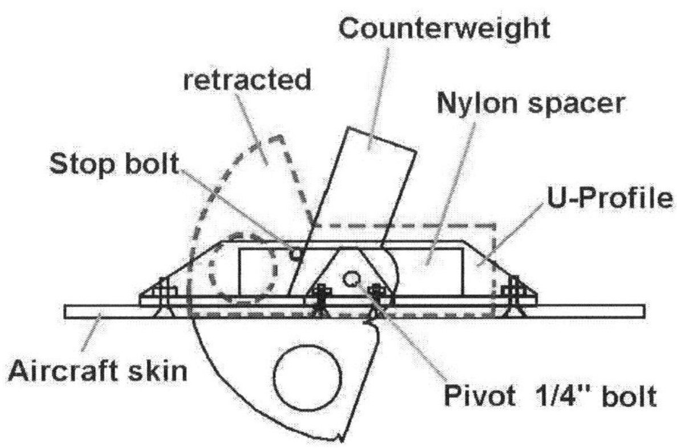

The left picture above shows the tie-down in the extended position. The tie-down assembly can be bolted directly to the skin using countersunk screws after you have reinforced the mounting location. An aluminum U-profile is cut to the desired length. The holes besides the pivot hole were drilled for weight reduction. A slot of the width of the tie-down plate is cut in the center of the U-bracket and the aircraft skin at the mounting location. The front edge can act as the stop in the extended position. Additionally you should provide a stop bolt to spread the loads. The pivot should be roughly located in the center. To increase the strength, the pivot hole is reinforced with L-brackets left and right. The actual tie-down ring plate can be made from aluminum or steel. To keep it centered and to act as guides, nylon blocks are used as spacers. The tie-down plate should be flush with the outer skin in the retracted position. This means it needs to protrude past the U-mounting bracket by the skin thickness of the airplane.

Static Ports

The static system supplies the instruments with static pressure. The instruments will only read correctly if the pressure in existence at the static port is very close to undisturbed static pressure. For this reason, most often the static ports are located on the aft fuselage. This means the boundary layer in that location can be quite thick. Flush static ports are affected by the boundary layer and are more likely to lead to wrong readings. It is better to have the static ports stick out a bit, about 0.08" to 1/8" (2-3 mm).

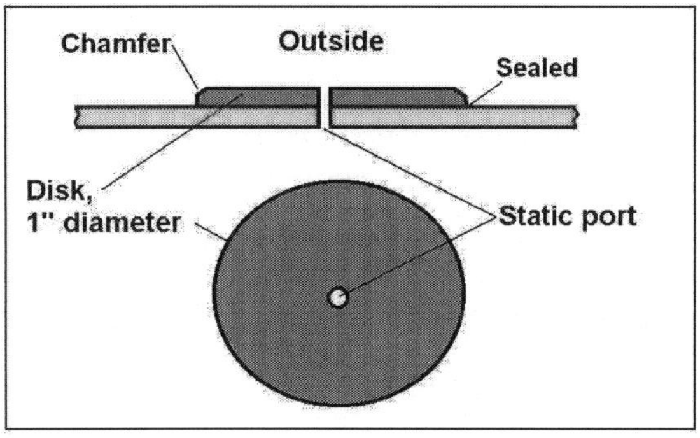

During the installation of the static ports, think ahead on how the static system check will be performed in the future. Seal the static port installation against the fuselage skin because during static check, a suction cup is placed over static port. If it is not sealed, a leak which is not in the static system, may show up between port and fuselage skin.

NACA Inlets

NACA shape inlets are the lowest drag inlets for airplanes for the amount of airflow they provide. This inlet style can be used for cabin ventilation or in the engine cowling to provide air to accessories. Various inlet shapes have been investigated by NACA and the one shown gave the best results. It is important that the corners except for the lip are not rounded in order to produce the vortices that direct air into the inlet. The dimensions are given in inches, but the whole inlet can be scaled up or down. For optimum effectiveness it is important to maintain the ratios given on the drawing. If you do not have the length available to fit a full sized NACA inlet, you can shorten it a bit and still get decent airflow.

If you use a NACA inlet for cabin ventilation, consider the location of the exhaust pipe and the rotation of the propeller. A propeller turning clockwise as seen from the cockpit will swirl the airflow clockwise around the fuselage. This means that the exhaust, discharged underneath the fuselage, will be drawn up the left side of the fuselage. If that is where your inlet is located, there is a danger that exhaust gas can enter the cabin through the inlet, especially at higher angles of attack.

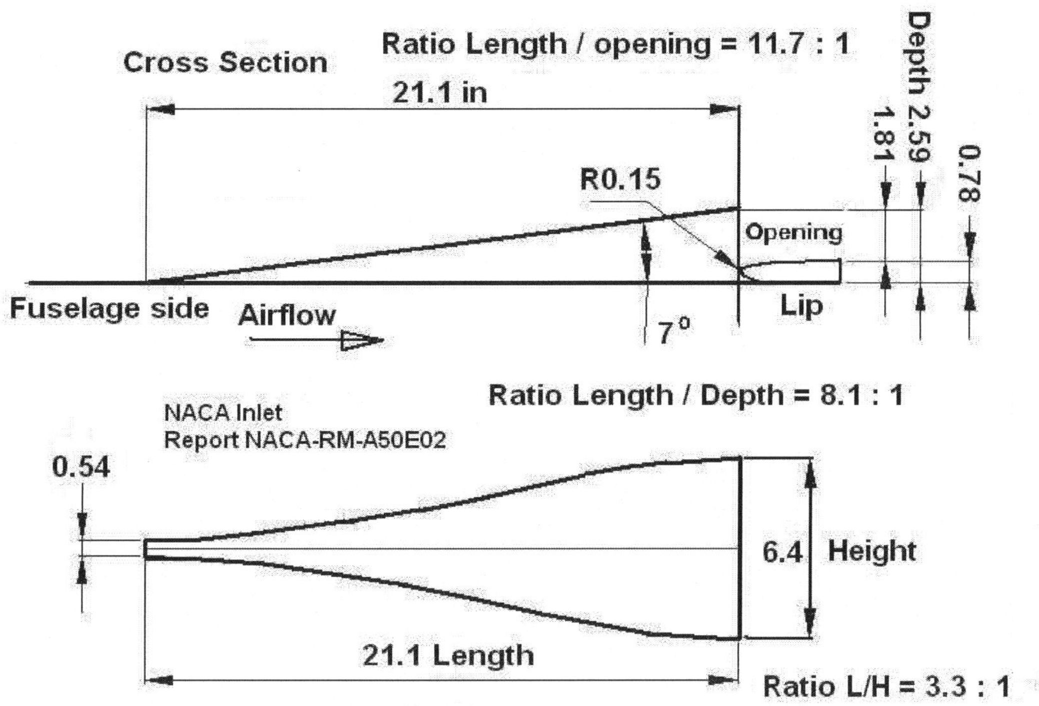

Assembly

Torque Witness Stripes

An easy way to remind yourself if a certain bolt was properly torqed is a torque witness stripe. This is a small mark in a bright color across the nut and onto the thread. The paint is brittle enough that it will crack or fall off if the nut has moved. You can either buy a small tube of the material or make your own from acrylic lacquer. When the lacquer is fresh, it is too thin. Leave some in an open jar until it has become goopy; then use it to mark the objects.

Torque Wrenches

Most structural bolt applications require a certain torque to keep things from coming apart under load. Torque is a force applied with an arm from the center of rotation. Torque is measured in a combined unit of force and distance. This can be inches times pounds (in-lbs) or ft-lbs if inches are replaced by feet. In the metric system Nm (Newton times Meter) is the most common unit. If for example the required torque is 12 inch-pounds, this is equivalent to the force of one pound applied at a distance of 12 inches. If you happen to have a wrench which is 12 inches long, you could hang a one pound weight from the end of it to give you the required torque. Fortunately there are easier ways to get there. Torque wrenches have a dial of some sort which lets you either read the torque directly or they can be set to a certain torque. Once that torque is reached, you can feel a click and changed resistance in the handle to indicate the torque you set has been reached. This by itself is quite straight forward. It only becomes more complicated if you need an extension. If an extension is used, the actual torque will be higher than what you read on the torque wrench. You can calculate what the torque wrench should read to have the proper torque at the bolt with the following procedure. You know what the torque T for the bolt must be. Measure the length of the torque wrench handle, A, and the length of the extension B. With the formula below, you can calculate the number that needs to be dialed in or read at the torque wrench. Remember that if the handle is extending the arm, the dialed torque must end up smaller than the required torque. The pictures show different kinds of extensions: a short open end, a set screw adapter and a hose clamp adapter.

T = required torque at the bolt D = torque dialed in on torque wrench

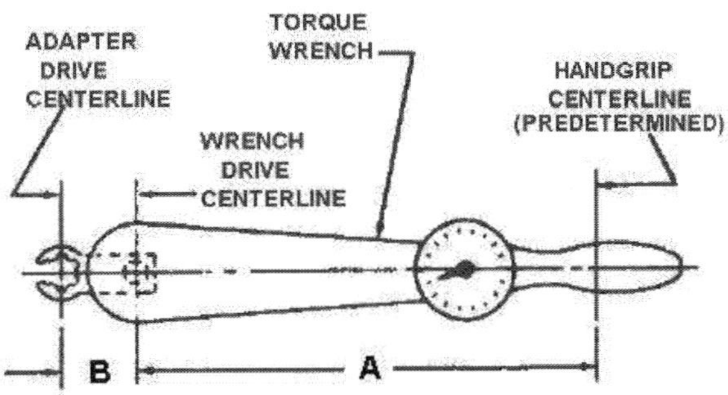

$$D = \frac{T \times A}{A + B}$$

Example:

A = 8 in, B = 1.5 in, T = 45 in-lbs
D = 45 x 8 / (8 +1.5) = 38 in-lbs

Since the calculation is based on the length, everything needs to be straight. If the extension is at 90 , no correction is necessary because the added length is 0. At any angle in between, you need to rerun the formula based on the new length.

Temporary Assembly

There will be many times when you need to temporarily hold parts together you can then affix them permanently. Unless the parts are match-drilled sheet metal where clecos can be used, a variety of materials can be useful. What they all have in common is that they take very little time to use. It starts with tape, regular or double-sided. You can use hot glue, which holds the parts as soon as it is cooled. Since the most temporary assembly fixtures are your hands, be careful not to burn them using this glue. If you do use this, get the 12 inch long glue sticks from a carpet store. Of course there is 1-minute and 5-minute epoxy; for example E-6000, which can be obtained in craft stores like Michaels. This can even be used for permanent work in an aircraft like applying a bond stud to affix aircraft wiring. These pictures show brackets temporarily assembled with hot glue.

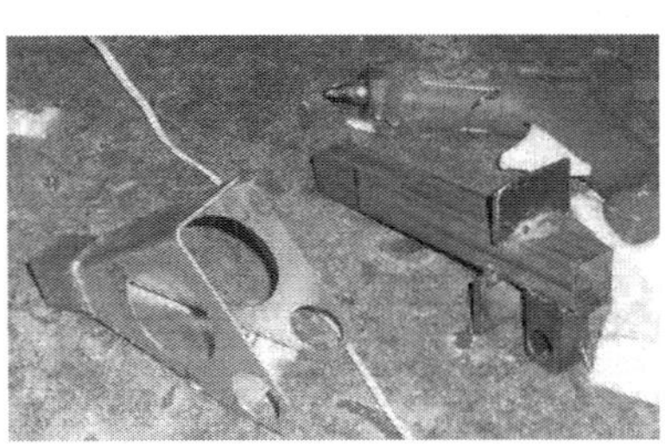 

Sliding Washers into Hard to Reach Places

What may work for you is a piece of wire. Slip the wire down through the hole, put the washer and the nut on. Hold your finger on the wire end and push the wire up through the hole and hold the nut and washer there. Use a small piece of rod to center the washer and nut, put the bolt down through the hole and turn.

Or you may find it helpful to use a piece of masking tape or duct tape wrapped around your index finger sticky side out. Do not wrap it all the way to the tip of your finger. Then put the washer about half on the tape - and half of the washer on the skin on the tip of your finger so you can feel it, then carefully feel for the bolt and push it on. If there is not enough space for your finger you can use a piece of sheet metal to stick the washer on.

When you build an assembly and you know there will be a hard to reach place where a washer is required, take a small dab of RTV (silicone) and stick the washer to the part and let it dry there. Then later all you have to worry about is getting the nut on the bolt.

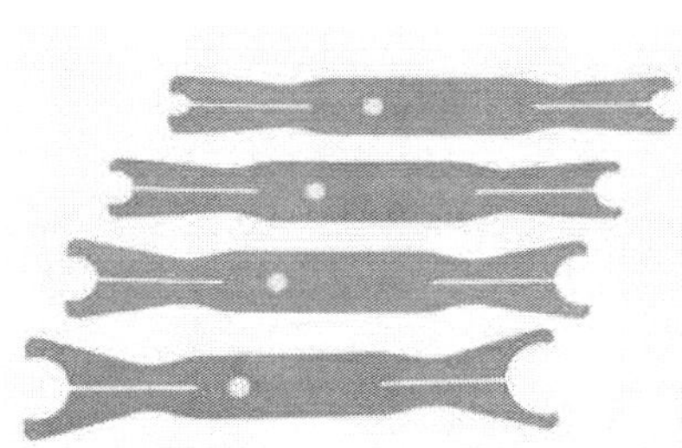

Tools you can buy are these washer holders (www.yardstore.com, part # 50199). They are made from sheet metal and will lightly clamp the washer while you insert it and push the bolt through. They hold a selection of eight different standard AN-sized washers.

Labels

Five or ten years from now, will you still remember what you installed behind the instrument panel and exactly where you put it? Probably not! It would be even harder for someone else to identify things if you should decide to sell your airplane at some point. When I bought an airplane someone else had built and not bothered to label things, I had a big problem when my electric trim quit working. The more than 16 small wires from the stick grip trim switches could not be identified without a wiring diagram and I had to replace the whole trim system. Get a pack of AVERY 8667 (clear) labels from the local Office Max and proceed to put Microsoft Word and a printer to good use. Make sure the surface is clean and apply the label in your choice of font and color. The components will be instantly identified such as the FLAP RELAY DECK, AMP TRANSDUCER, LIGHT DIMMER BUSS, ROLL TRIM RELAY, PITCH TRIM SPEED GOVERNOR etc. Wires can be labeled in a similar manner.

Leveling the Plane

When you build your airplane, think ahead at the task of leveling and rigging. It helps if the airplane has a straight surface somewhere which is parallel to the waterline. Look up in the plans how the waterline is defined. Especially curvy composite planes hardly have a straight line on them where adding something straight as an aid to place a level later on can make that job a lot easier. Canopy rails are useful, but may not be parallel to the waterline. Make sure you know what their angle is. To make the job easier the second time, you can establish a reference of your own choice. You need a high spot inside the fuselage from which you can hang a string with a plumb bob. With the airplane leveled longitudinally and laterally, the tip of the plumb bob will point to a certain spot on the floor. Mark this spot permanently, for example by bonding a small washer to it. Later, when you need to level the plane, adjust it until the plumb bob points to the center of the washer. The further apart the upper support of the plumb bob string is from the washer, the more accurate will be this method.

Wing Incidence

To assure that your airplane flies straight, it is important to have the correct incidence at all wing stations. Most airplanes have an untwisted wing, which means the incidence is the same at all wing stations. Some designs use washout (reduced incidence) towards the wing tip to improve stall characteristics. To check the incidence on the finished airplane, you can build a simple tool. A straight bar with a vertical stop for the leading edge is drilled in two places for threaded rods. The threaded rods are used to level the tool at a particular wing station. Thus set up, move the tool to the same wing station at the opposite wing. The incidence should be the same. If your wing has no washout and has constant chord over the span, you can use the same setup for the whole wing. If you have a tapered wing, you can only compare the same wing stations from the left wing to the right wing with the same setup. To compare apples to apples on a tapered wing, or in other words to use this tool without adjustment in height, you need to move the threaded rods to the same positions chordwise (for example at 20% chord and 80% chord) over the span because the chord varies. For this case you should provide slots rather than holes for the threaded rods. To measure the angles of a twisted wing, use an inclinometer or Smart Level.

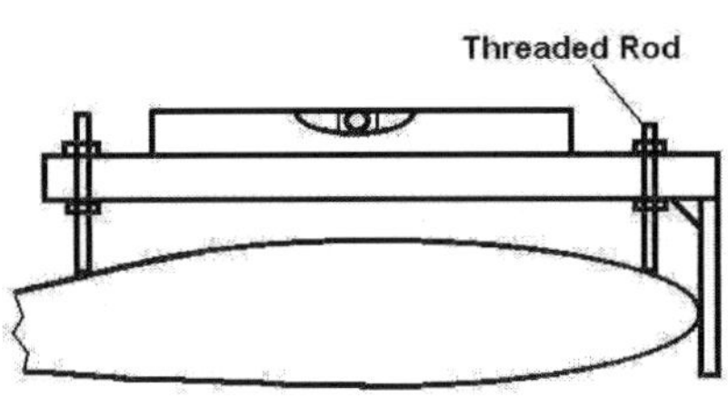

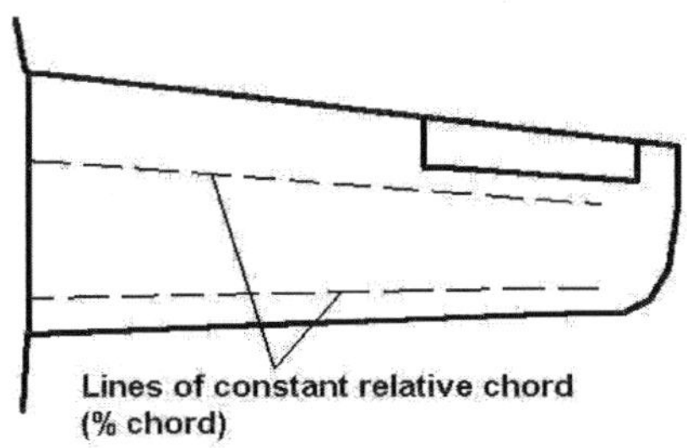

Final Assembly Check

During the final assembly, you should check the overall geometry of the airplane and its symmetry. Good points to measure to and from are wing tip to horizontal stabilizer tip left side to compare with the same measurement on the right side. Measure from the airplane nose to each wing tip and compare. If the engine is installed at an angle to better counter torque and prop slipstream, this distance may not be equal left to right. Measure from the tip of the vertical tail to both tips of the horizontal tail. Check the horizontal tail incidence. On most airplanes, the horizontal tail incidence is a few degrees less (leading edge down) than the wing incidence.

Threading the Frayed End

To thread the frayed end of a rope, roving, cable or similar object through a small hole, wrap it with tape, spiral wrap or a drinking straw cut into a barber-pole spiral at the end. Thread it through the hole and remove the threading aid. Or dip the end of thread or rope in some Super Glue; let it dry; push it through the hole and cut off the glued portion.

Door Gaskets

If your airplane has gull wing doors, take a look at the hinges. If you have not flown the airplane yet, you probably don't know yet if water will drip into the cabin if it rains. Chances are good that it will drip, unless you have an effective seal. The water of course will run to the lowest point. All along the edge and at the door hinges, there is nothing to stop the water from going into the cabin without a seal. If the lip of the door flange was curved up, it would form a trough in which the water could be routed downwards along the sides, if it is a small amount of water.

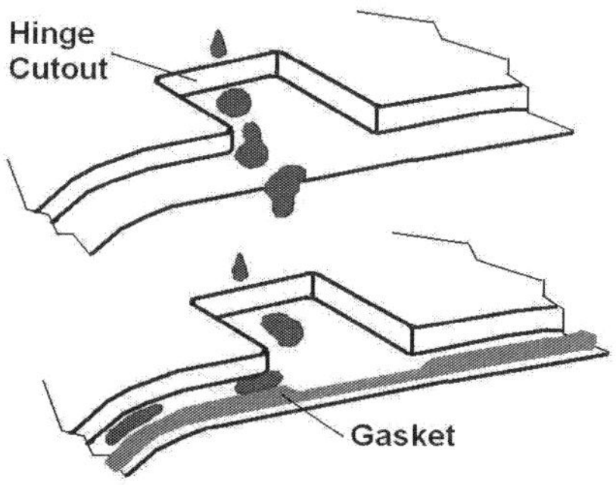

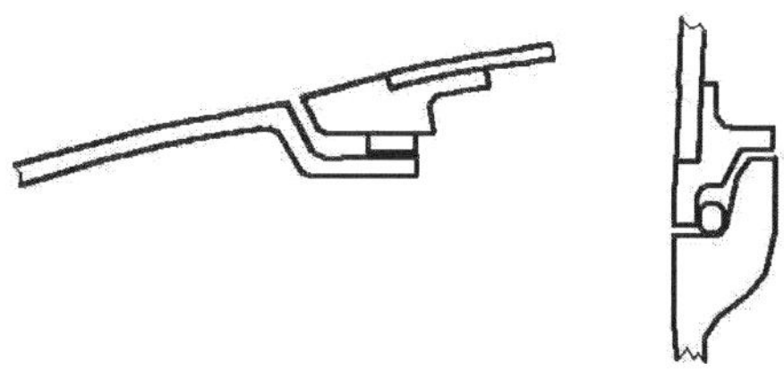

It is better to glue a gasket along the edge of the door frame, as shown in this picture. At the hinge is a low spot, there the gasket is compressed by the hinge.

Typically, rubber seals come in 3/8 inch and 1/4 inch thickness, 1/2 inch wide, with adhesive backing. You put them on, and then close the door and let it set for at least 24 hours. The first door closing is difficult, because you have to compress the foam. The foam then takes a set, and the subsequent closings are easy. If you have a 1/4 inch gap, you should use 3/8 inch foam. In the area of the door hinges, cut the thickness down to ~3/16 inch and close the door on it to compress it down to ~1/16 inch. When you put it on the sides of the door, you may have to do it in stages, because you might not get the door closed if all the foam were in place at once. Once you have it all installed, conduct an experiment on your door seals. Close the doors and pour a glass of water on the door hinges, both front and rear. If your seals work well, there should be a small space behind the seal that allows the water to drain off down the side and exit on the outside. The cockpit should stay dry. Other examples for door or canopy frame shapes are shown in the sketch on the right.

Vent Holes and Drain Holes

Any bonded airplane is at risk of having completely enclosed airspaces in its structural components. It must be avoided to have any airtight compartments in an airplane. Because of pressure changes with altitude changes, a sealed compartment can be subjected to severe loads. The higher the airplane climbs, the more the trapped air tries to expand and push against the walls until in the worst case a bonded seam may fail. For this reason, drill at least one small hole in each rib or bulkhead to allow the pressure to equalize.

Air leaks from gaps and seams increase the drag of an airplane. But if the airplane is too well sealed, undesirable liquid is unable to drain. I have seen it several times that considerable amounts of water accumulated in low spots of an airframe after the airplane had sat outside in a rain shower. Especially if water is allowed to accumulate in the tail, it can lead to a dangerous shift in CG. Other effects are increased corrosion or damage if the water freezes. Every compartment with a low spot should have a drain hole, either to the outside or to another compartment which has a drain hole to the exterior.

Slip Ring Brushes

If you have electric propeller pitch adjustment, the motor in the hub is supplied with power through brushes and a slip ring. The same arrangement can be found if you have de-ice boots on the propeller. The carbon brushes are spring loaded and may not be secured other than by the slip ring itself when the propeller is installed. To keep them in the brush block holder and prevent them from getting damaged when the propeller is installed, secure them with a tie-wrap. The tie-wrap can be cut once the prop is installed.

Sizing Access Holes

The first thing to determine is how large the hole needs to be for each location. The answer is: it depends. It depends on the purpose of the hole. Is it there just to inspect things at the annual or will you need to get tools in there and work? How big are your hands? Maybe you need a larger hole. Before cutting any holes in aluminum, fit a piece of cardboard to the wing ribs then cut what you think are the right sized inspection holes in the cardboard. Then, using some actual tools, try to reach in and try working. The hole in the cardboard may need to be gradually enlarged until you can perform whatever operation you are trying to do. Now using that information, you can make your inspection holes.

Piano Hinges

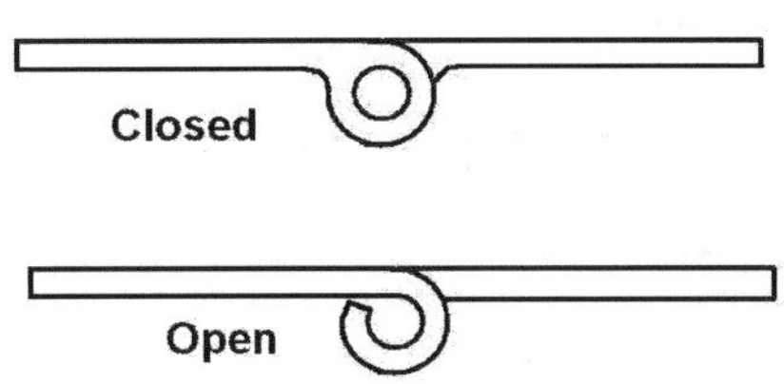

There are two basic hinge styles, open and closed. The open version is cheaper, but not as strong as the closed version, which is made from an extrusion. The open hinge should only be used for non-structural applications such as small access doors.

A piano hinge may be riveted or screwed on. If you rivet it on, the two components can only be separated by pulling the hinge pin. It may make sense in some cases to rivet only one side and use screws on the other side.

The hinge pin needs to be secured at the ends to keep it from working its way out of the hinge. The most secure way is to make the pin longer than the hinge and bend the ends. This will of course prevent the removal of the pin, should that ever become necessary. A second method involves a pin slightly shorter than the hinge and pinching the ends of the hinge closed. This also makes removal of the pin difficult. The third and fairly common method uses a slightly shorter pin. At the ends of the hinge, drill a small hole and insert a piece of safety wire or a cotter pin. The safety wire should not be too thin and the end of the hinge pin should not be pointy. I have seen this situation, where the hinge pin in a somewhat worn out hinge managed to get past the safety wire and fall out completely. Of course you can use a combination of the described methods. Having one end of the pin bent and sticking out makes it much easier to remove it. If the configuration allows, you can secure the bent end of the pin with safety wire as shown in the following picture on the right.

Some piano hinges open almost 180º in one direction, but very little in the other direction. Keep this in mind when you install them. If the door opens to the inside, as shown in the center of the left picture, the angle how far it opens may be limited by the interference of the fasteners. Should the attached piece move in both directions as it is typically required for control surfaces, the hinge must be attached to surfaces which are near right angle to each other as shown in the center picture.

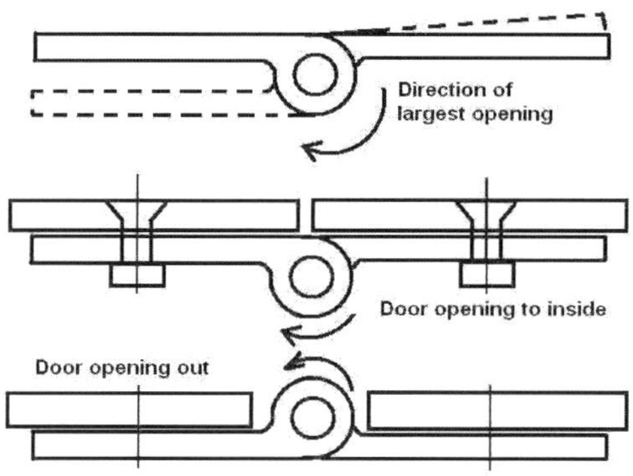

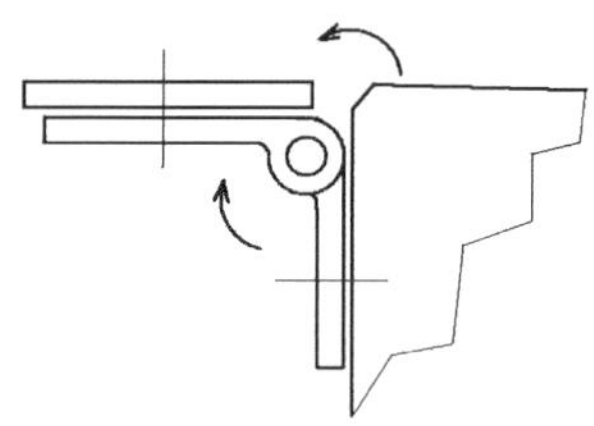

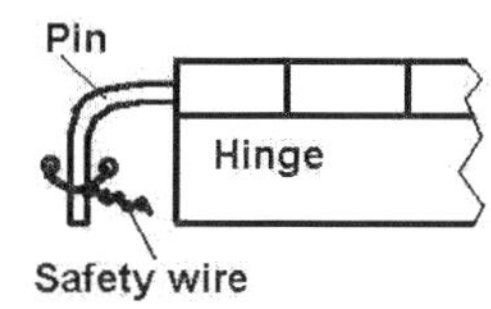

Camlocks

Camlocks are convenient fasteners for components which need to be opened or removed frequently. They are operated with a screwdriver and only require a quarter turn to unlock. They can be mounted flush with the surface, which reduces drag. There are a few things you should be aware of when you consider them for your airplane. They are spring loaded, which means they will hold two pieces together only by the force of a spring and not with a thread like a screw. If there is a lot of tension between the two parts, the spring will allow them to move apart slightly until it bottoms out. In most applications this is not an issue. Camlock fasteners work best in shear. The main drawback is that to install the grommet, the hole needs to be slightly oversized.

These pictures show a stud, receptacle, grommet, snap ring and Camlock pliers.

When the grommet is inserted in the hole, a snap ring on the inside keeps in place. This allows the parts attached with camlocks to slightly move radially. When you drill a hole for a grommet, make it only large enough for the grommet to just about fit through. The snap ring is not wide enough to hold it in an oversized hole. Unless you use an adjustable version, the width of the grommet needs to be selected to match the thickness of the material it is used in. A receptacle is attached with rivets or small screws to the other side. The studs are available in different sizes to fit different material thickness. You need to combine the proper grommet and stud sizes for your particular application. Typical stud sizes are number 3 to number 5 for sheet metal. Composite is usually thicker and you may need to range up to number 6 or 8. The receptacles are all the same size. There are a few different Camlock styles available. I would recommend using only stainless steel Camlocks on airplanes, the plated metal ones will look rusty sooner or later. A stud is inserted into a grommet with special pliers. If you have to install a lot of them, it is worth getting the pliers. For a few, you can get by with needle-nose pliers and some fumbling. Or you can buy a cheap pair of pliers and alter them with a Dremel tool to emulate the official pliers. The stud has to be compressed to fit through the hole in the grommet. Once installed, it will not fall out on its own. The stud is available for Philips screw driver (shown above) or slotted for a flat blade screw driver. I only use the Philips screw style studs. When you lock a camlock, you have to push down on it. A flat blade screw driver easily slides out of the slot during this operation and the result is a nasty scratch in your paint.

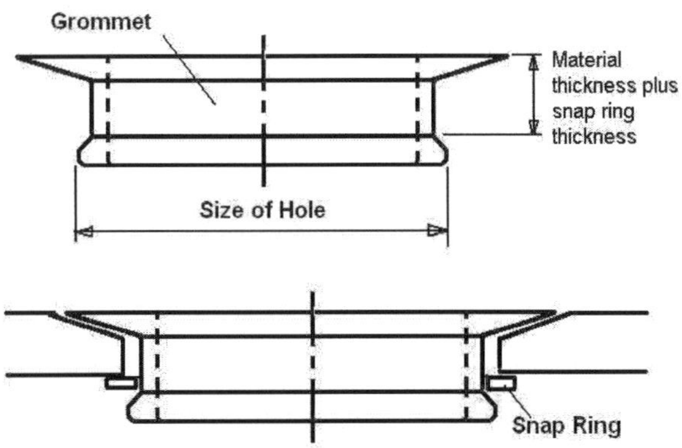

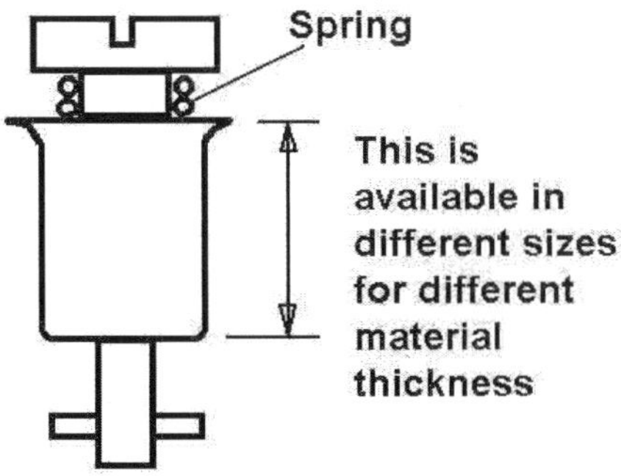

Hinged Instrument Panel

If you ever had to get to the backside of an instrument panel, you know how much trouble it is to either remove it or lie on your back in the foot well reaching up to get to it. One builder has come up with a unique solution to this problem. He fabricated aligned hinge pins on both sides an in the middle of the panel using the original instrument panel from the kit. He has setup the panel so that the avionics will be in a fixed panel in the middle with the remainder of the panel modified to fold down. The pictures show the panel in the upright and the folded down position. This setup requires a bracket to hold the panel in place in the upright position, which is not shown here. Once the instruments are installed, the tubing and wires have to be of a length and arrangement which will not prevent the folding down.

Instrument Panel Lower Edge

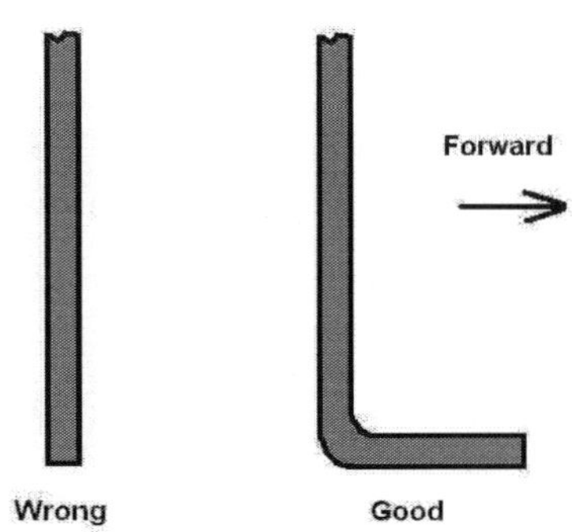

The lower edge of your instrument panel should never be a straight edge. Instead, it should be an L-shape. A straight edge will act like a knife and can severely injure you in an accident. The L-shape provides a much larger surface area and has the added benefit of structurally stiffening the panel. This may allow the reduction of attach points and reduce instrument panel vibration.

Storage Space

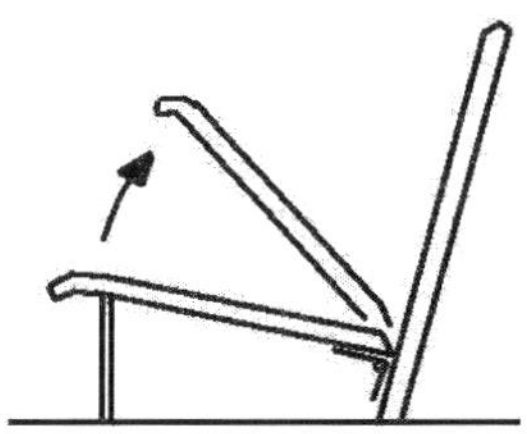

To increase the storage space in your airplane, look under the seat. Depending on the design of the seat, you may be able to install a hinge to the seat bottom to be able to lift it up and use the space underneath to store things. Mine has its toolbox under the seat. The seat bottom is at the same time the lid of the tool box. Your or the passenger's weight will hold the "lid" in place, but you should use velcro as well in case you roll the plane inadvertently and no one is in the second seat.

Trim to Fit

There are many things which need to be trimmed and fitted in place. The basic problem is you have a surface of irregular shape and are trying to cut and fit a rib, bracket or something else to match that contour. Use a piece of paper or cardboard roughly cut to shape as the starting point. Hold it in place, mark up where it does not yet fit and trim it. This can go back and forth several times until there is a reasonable fit. Transfer the pattern to the real part and cut it out. Most likely there will be some fine tuning needed to reduce the gaps. By shining a light on the part from one side, you can easily see where exactly it needs to be trimmed. The gaps will show up as bright patches. It stays dark where the part touches the surface, as you can see in this picture.

To save time, you can use a piece of cardboard cut roughly to size and modeling clay. Protect the structure with tape or a sheet of plastic. Hold the cardboard template in place and push a bead of clay around the perimeter to fill the gaps. After removal this retains the exact contour you are trying to create.

Duplicating Parts

If you need to cut several parts of a certain shape and you want them all to have the same shape, use standard woodworking router practices. First you need to make the perfect master part. Then use a band saw or a jig saw to rough cut next piece and tape the two together with double sticky tape. Use a router table to make as many duplicates as you want. Do not clamp the parts together if you use tape to hold them together, they may become very difficult to separate. A template router bit (right picture) can be used to duplicate a part using only one template. It has a bearing which follows the contour of the template while the bit cuts the piece clamped to the template to the exact same shape.

A different technique requires you to make two templates and clamp the other rough-cut pieces between them. Now you can sand or file the whole block down to the contour of the pattern. Just make sure the templates are aligned to each other and there is no twist. This method can also be used to make shapes of tapered size if one template is smaller than the other one.

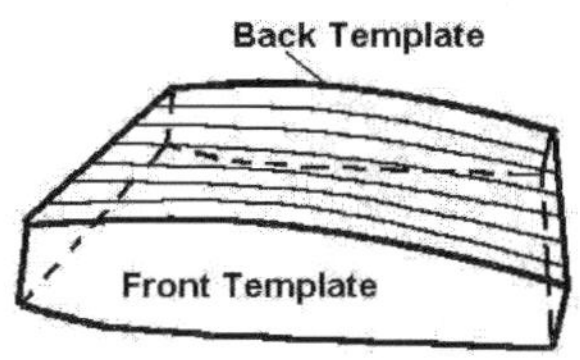

Locating Blind Holes

If you need to match drill a hole in a sheet that covers the one with the hole already in it, a hole finder is the most useful tool. Hole finders are designed for one particular hole size only, so you need to have several sizes available for the most common hole diameters. There is another way to locate blind holes; if you do not have a hole finder handy, try this. Place a strip of masking tape on the under side of the part you want to make the matching hole in. Then put the part in place over the hole and press down hard. When you take the part off and look carefully at the masking tape, you may be able to see where the hole needs to be from the impression in the masking tape. This method may depend on a lot of little things, such as if the existing hole has a rough edge, how hard you can press, how big the hole is, how thick the masking tape is, and how thick the metal is.

Inserting Bolts in Tight Spaces

As anyone who has mounted control surfaces knows, getting the bolts into the bearings of control surfaces can drive you crazy. Here is a tool that helps get the bolts started in a very confined area. It consists of two pieces of scrap stock that are formed as shown and riveted together. The slot catches the bolt and the two pieces are formed so the bolt head is held in the slot by friction.

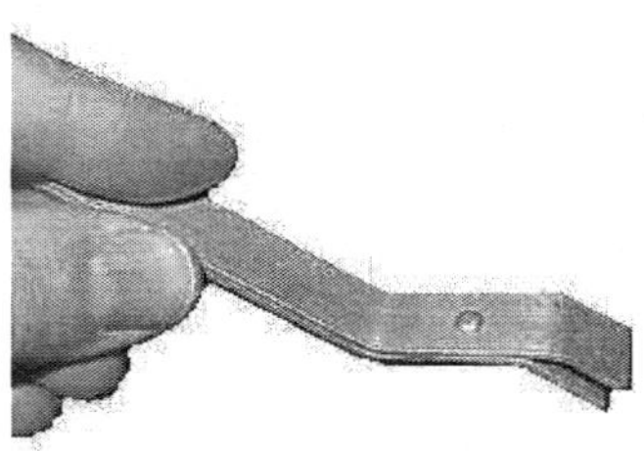

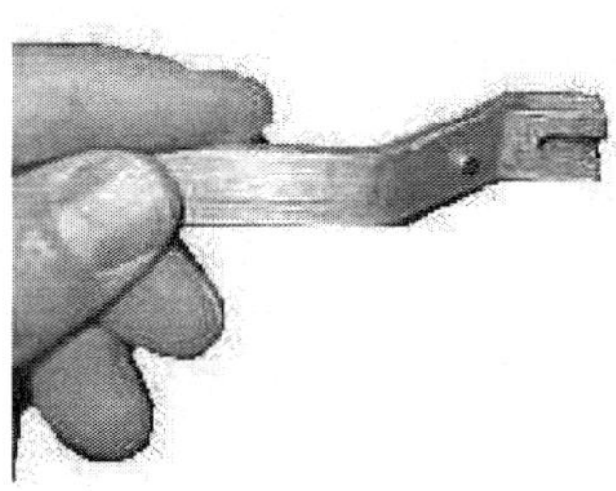

Drill Guide

Have you ever tried to drill a hole through a wide board with the intention to have it come out the other side exactly were it is supposed to? And just how successful were you? Using a drill guide will help you get it right every time. With the entry point defined by the drill bit, it lines up the exit point using a pin. The simplest way is to take a block of some sufficiently thick and stiff material, drill a hole all the way through with a long drill bit and cut out the center. The width of the cutout should just match the thickness of the part you are trying to drill (right picture).

In the left picture you see a different version of a drill guide. The pin is on the left, next there is the body, made from soldered copper water pipe. The long drill bit is on top. Of course there are many ways of making this tool. Take a piece of 12 inches long steel tubing and run it through the copper from one end of the tool to the other with a little bit sticking out both ends. Bond the steel tube inside the copper tubing by filling the gaps with glue. Determine the width of the plank to be drilled and cut out a section of the steel tubing accordingly. Now the drill bit will be guided absolutely straight to the exact point it has to hit on the other side of the plank. No guesswork involved. To drill a different size hole, remove the steel tube by heating it sufficiently for the glue to soften and pull it out. Bond a different sized tube into the copper frame to repeat the process.

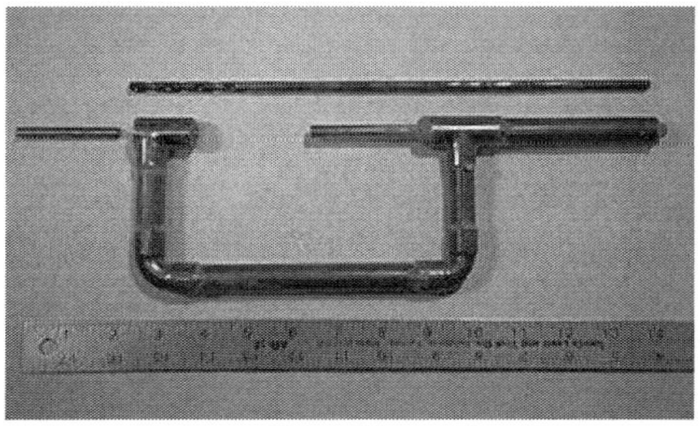

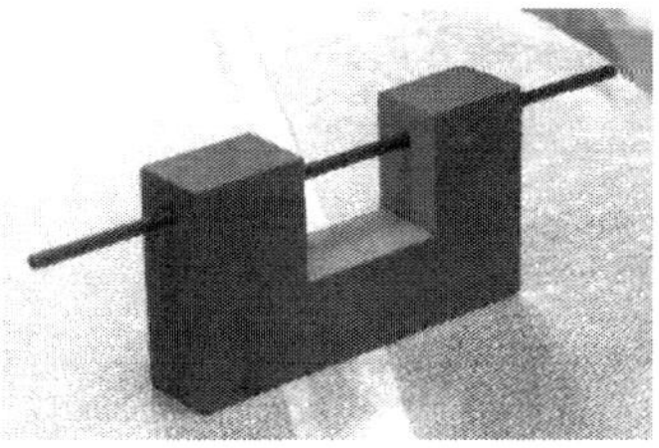

Easy Paint for Small Parts

Spray-can paint is often under-rated. I have used it for a number of small parts, especially inside the airplane. With a bit of practice, you can get good looking results. Of course everyone will have their own preferences as to which brand they like best. For your panel or other interior parts for example, a nice gray color often works well. The best paint for this is NAPA 7250 Iron Cast. The paint has an extraordinary characteristic in that touch-ups look as good as the original. So when you scratch the piece or need to cut a few more holes, touch it up with the can and the result still looks like the original. Parts requiring a smooth, fairly hard surface can be painted with high-temperature automotive paint designed for engine compartments as long as the part does not come into contact with fuel. It costs a lot less to buy one spray can for one color you need in small quantities than to buy a can of two-component urethane paint.

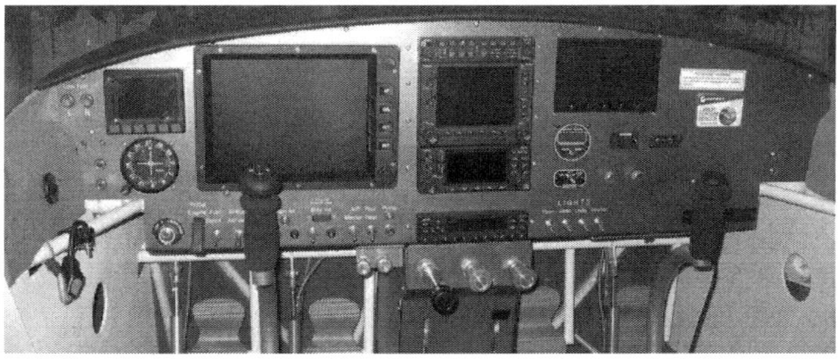

Powder Coating

Powder coatings resist cracking and peeling and are resistant against abrasion, corrosion and chemicals. The performance characteristics of powder coating match or exceed the characteristics achieved from conventional solvent-based coatings (paint). As a homebuilder, you can either get set up to do powder coating yourself or send out your parts to a professional shop to get them powder coated. Unless you have a lot of parts or want to use this process frequently, the latter is a lot less hassle. The powder coating process uses a solvent free dry mix of plastic resins, pigments and fillers that melt and fuse together when heated. The solid particles of coating are electrostatically charged in a spray gun and are carried by low velocity air to the surface of the piece to be coated. Electrostatic charge holds the powder particles in place while the part is cured in a 400ºF (200ºC) oven. The heat of the oven causes a chemical reaction to occur and the powder to cure, creating a highly durable finish. Powder coating is an environmentally friendly process unlike solvent-based wet paint systems. Any metal object that can hold an electrostatic charge and withstand the heat of the curing process can be powder coated.

There are some things you need to observe even if you send your parts out for powder coating. If there are areas that need to stay free of coating such as holes and threads, your masking tape or plugs need to be able to withstand the high temperature in the oven. Silicone works well. For the same reason, the use of plastic and resin body fillers is problematic. Often the powder will not bind to a filler's surface causing bubbles in the coating. To be safe, fill any depressions with a torch using brass or welding rod as fill.

Preparation of the part for the powder coating process is similar to the preparation for paint. Remove all oil, metal oxides, welding scales etc. Sandblasting is a suitable treatment for steel. Aluminum must be chemically treated (etched) for the coating to adhere. The most common way of applying the powder coating to metal objects is to spray the powder using an electrostatic gun. The gun imparts a positive electric charge on the powder, which is then sprayed towards the object, which is grounded. The object is then heated, and the powder melts into a uniform film, and is then cooled to form a hard coating. While it is relatively easy to apply thick coatings which have smooth, texture-free surfaces, it is not as easy to apply smooth thin films. As the film thickness is reduced, the film becomes more and more orange-peeled due to the particle size and glass transition temperature of the powder.

When it becomes unusually difficult to get powder onto the part or into the inside corners (Faraday cage effect), there are three common things that should be checked:

- Grounding
- Powder flow rate
- Voltage

Grounding may be lost from excess build-up of coating on the hangers. Measure the resistance from the part to the hanger with an ohmmeter. If the resistance is greater than 1 megaohm, the path to ground is not adequate. Find the insulated component and clean it to make good contact.

Excessive powder flow rates will cause a reduction in charging efficiency. Check the gauges and reduce the flow rate.

The voltage at the gun tip needs to create a field of high potential. Check the voltage at the gun tip to make sure that it is compatible with the gauge.

The Faraday cage effect describes the problem with getting the powder to adhere to the surface of an inside corner. This is the result of resistance created by the presence of an electrostatic force. Faraday cage effect can be reduced by limiting the current draw of the gun, by using the correct nozzle for the task, and by control of the flow rates and gun-to-target distance.

Pin holing can have several different reasons. For example if the alloy material is porous, the air in the pore is expanding when the parts are heated in the oven. The air then escapes into the powder film when the powder is in the liquid or melted state and as the powder solidifies during the curing process the air then becomes trapped in the film and forms pin holes. To help reduce pin holes, it is recommend to preheat the parts, bring them up to the cure temperature and hold it for at least 10 minutes, so the metal alloy will have a chance to heat up. This will help to chase the air from the pores. Coat the parts while they are still warm; then subject them again to the cure temperature.

If you use more than one coat (after the first one is cured), there is less attraction on the second coat. Try to use more air and fogging. Spraying while the part is hot should also help.

Primer for Aluminum Parts

Prime everything! I have seen enough corrosion on metal airplanes to know that it is important. And it develops everywhere, not just where you can easily inspect and repair it! Prime all parts after you are finished with all cutting, drilling and deburring operations, prior to assembly. Do you really want to take a chance and save some time by omitting this process? Only if you live in a desert and never fly anywhere else, you can skip this chapter. The description in this paragraph is typical for the etch-alodine-prime process. Depending on which primer you select, it may be slightly different.

3M Alumiprep is used as the first step. This is the acid etch process. It will remove any grease, oil and dirt from the part and prepare the surface for the adhesion of the primer. Mix the Alumiprep with water per the manufacturer's specification, then wash the parts well with the solution. The part markings or writing on a lot of the parts will come off during this step. This will help you know you have done a good job. Rinse the parts with water, and let them dry. They should now be dull instead of shiny. Do not touch the parts with bare hands after you cleaned them. Make sure you are wearing gloves to both protect your hands and to keep your hand oil off of the parts.

Next the parts are alodined. This is the step that turns the aluminum parts gold-looking. Actually, the gold color comes from a dye that is added - it has nothing to do with the chemical reaction. It is added only so you can see the progress and tell which parts have been treated. You can also get clear alodine which will not change the color of the part. The alodine solution should be fresh. As the alodine gets older, the dye does not work as well. Follow the directions for how long to leave it on the parts, do not rely only on the color. You can completely immerse parts in the alodine, or use a spray bottle to spray the larger parts. After the required time, rinse with water and let them dry. Depending on the exact aluminum alloy of the parts, some will end up more gold than others.

The two previous steps remove the oxide layer and treat the material surface to promote adhesion of the primer and delay the formation of a new oxide layer until you can apply the primer. You should prime within 24 hours of the etch and alodine process or repeat it.

Last is the primer. There are many suitable primers out there. Here are some considerations for selecting a primer.

- Is it available locally and has good support and data sheets? Saving on shipping cost can be a big deal.
- It is a corrosion inhibiter (zinc chromate), or just a moisture barrier?
- Does it contain fillers to allow sanding? Primers without fillers are lighter weight and eliminate the chance of moisture absorption by the filler.
- Does it require sanding or roughing up of the aluminum substrate or just cleaning?
- Is the primer self-etching?

Two-part epoxy primer is good, as are zinc chromate primer, zinc oxide and lacquer primer. When you select a primer, consider the topcoat that will be used over it. Not all combinations of primer and top coat will work. Do not forget the dimples on the outside of the skins. Those need priming too. A lot of the corrosion shows up around rivets. Dissimilar metal and the fact that the rivet hole cuts through the alclad probably contribute to the faster corrosion in this area.

Changing Tires

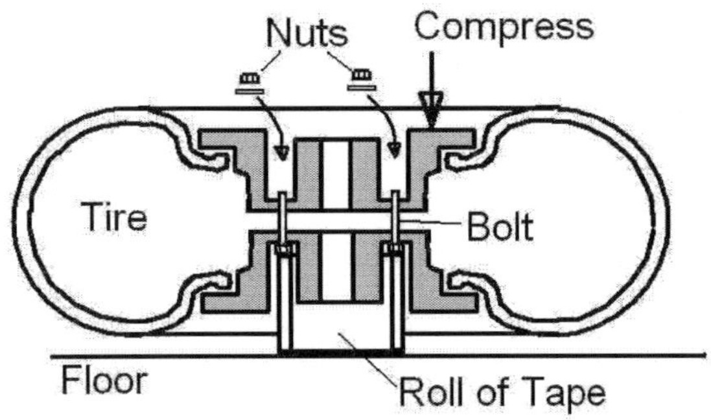

Two-piece hubs are bolted together with the tube and tire in between them. It can be a challenge to assemble the halves by yourself because of the springiness of the tire. The job becomes very simple with the help of a roll of tape. Place the roll of tape or similar object underneath the bottom half of the hub and stack the rest on top of it. It will hold the bolts in and allows you to get the nuts on while you compress the assembly.

Fabric Covering

Rib Lacing

The fabric on most airplanes needs to be laced to the ribs to keep the air loads at higher speed from pulling the fabric off the surface. You will find that shining a light up through from the bottom of the wing or control surface helps quite a bit in locating the long needle as it gets pushed through from top to bottom.

Avoid Fraying Fabric

The fabric tends to fray when you cut it or otherwise damage the fibers. One method of minimizing the fraying along a cut is to mark up the line along which the cut will be and brush a little glue on it. If you need holes in the fabric for rivets or drain holes, use the tip of a hot soldering iron. It will melt rather than cut the fibers although this may leave a small ridge around the hole from the melted material.

Avoiding Glue Stains and Wrinkles

When the fabric is glued around the tube of a flying surface, it is easy to stain the fabric on the other side. Here you can see the fabric glued around the tubing on one side. A piece of cardboard is a way of keeping glue off the inside of the fabric when you attach the fabric.

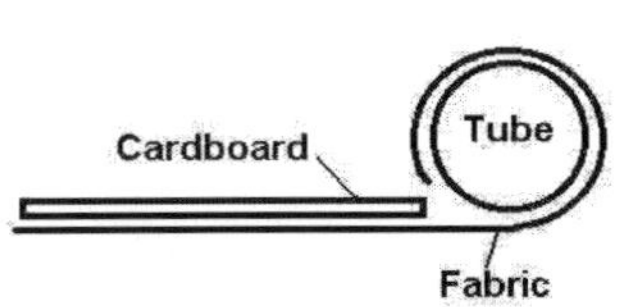

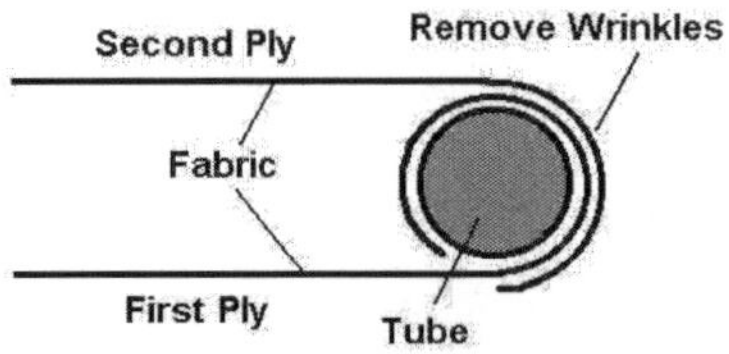

Once the fabric is glued to a tube or other structure on one side, the next step is to use the iron at about 225 F to smooth down the glued fabric. You need to smooth this down so when you glue the other side you are attaching to a good surface. It is hard to smooth things out after you have more than one layer.

New Fabric Covering Material

Have you ever thought how nice it would be to cover an airplane like a model airplane, with foil ironed on as the only step to create a colorful, glossy finish? Actually you can do that. There is a new material available called Oratex 600 which reduces the covering process to one step, the covering only. No dope, paint or hours spent sanding. This is the easiest covering process I have used so far, and it really saves a lot of time.

Oratex 600 is a coated polyester fabric of about 110 g/m^2 (3.2 oz/yd^2). This weight includes paint and UV protection. It is airtight and available in a number of different colors. It can be painted if that is desired. Its tensile strength is 310 N per 25 mm (70 lb per in) in fill direction and 290 N per 25 mm (65 lb per in) in warp direction. It is non-flammable and can be tightened with heat at temperatures up to 170ºC (338ºF). Its shrinkage is 5% to 7%. The melting point is 250ºC (482ºF). The paint, which is part of the fabric, it quite thin, which saves weight but it also lets you see a bit of the surface underneath, at least if there are darker marks underneath.

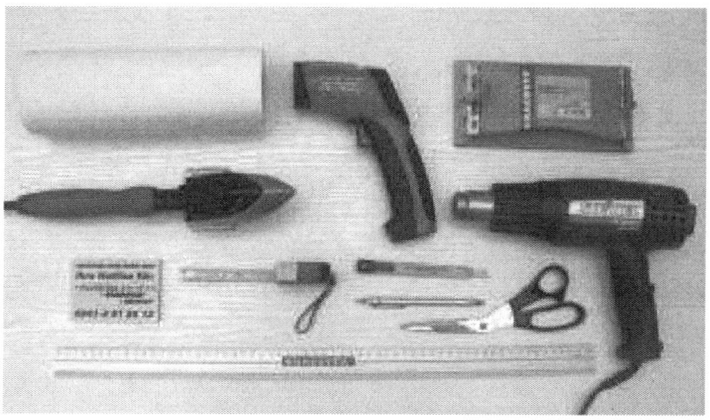

The Oratex 600 fabric is applied with heat sensitive glue. The glue is spread on the part and on the fabric. Before further work, the two-component glue must dry completely. These are the tools needed to cover an airplane with Oratex 600:

- Iron
- Scissors and ruler
- Brush or spray gun for the glue
- Tape Measure
- Felt pad, rags
- Heat gun
- Sharp knife
- Sanding block
- Thermometer

The materials needed for a covering job are:

- Oratex 600 fabric
- Dispersion hot glue
- Reinforcing tape (if needed)
- Thinner (may be needed for spraying)
- Silicone paper

The preparation of the part for covering is no different from a regular covering job. The part must be clean, free of sharp corners and degreased and primed or painted if the material is metal. Spread the fabric on a flat surface, lay the part on top of it and mark the fabric with a pencil where the fabric will be glued to the part. Apply the glue to the part on all surfaces where the fabric will be glued on. This can be done with a brush or sprayed on. Apply glue to the inside surface of the fabric where it will be glued to the part. Let it dry completely. To bond the fabric to the part, calibrate your iron to a temperature of 90ºC (194ºF). Attach the fabric to one side of the part by ironing it on. If stitching is required at ribs, do this in the same manner as with a conventional covering job. The tape to protect the stitches is glued and ironed on as well. Continue to iron the fabric on until it is attached everywhere.

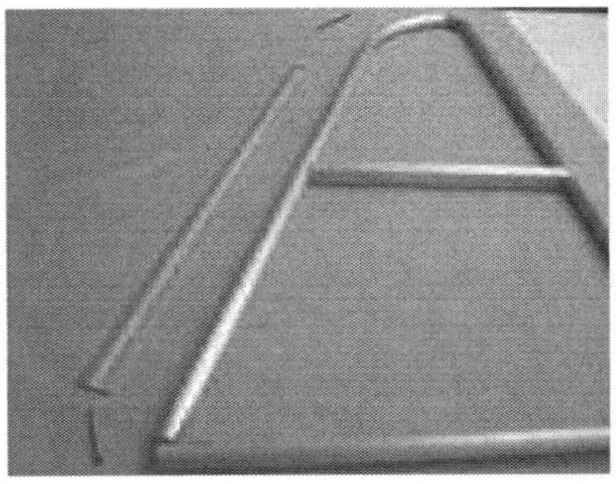
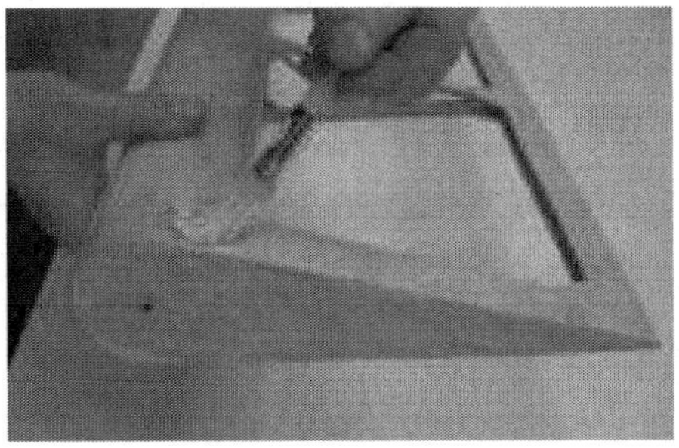

There may initially be wrinkles in the fabric where it is pulled around round corners but with the application of heat these should disappear. Once the fabric is completely attached, it is probably not completely smooth yet. Shrink the fabric with the heat gun until it is tight but do not exceed 200 ºC (392ºF). The process is the same whether it is a control surface, wing or fuselage. It can be stretched around tubes and curved surfaces.

The fabric is suitable for airplanes in the LSA weight category with a top speed of about 120 knots or less. Currently it is only available in Germany from the Lanitz-Prena Folien Factory (http://www.oracover.de). It costs about 31 Euro per square yard. This may seem fairly expensive, but keep in mind that this includes the paint and saves a lot of work. None of the materials used are hazardous or generate fumes and can be applied with without special personal protection. I covered a rudder on the carpeted floor of my living room, no mess to clean up.

Powerplant

Protection of the Inside of the Cowling

Engines always leak or spill something. And the cowling is the structure which catches it. Because most cowlings are composite, they will to a certain degree soak up the fluids and get stained. Make your cowling easy to clean and oil-proof by coating the inside. Whatever you use should be able to handle the heat from the engine. Normal paint may discolor or, in the worst case, blister. There are high temperature automotive paints available which are very suitable. Select a light or reflective color to keep the cowling temperature down. Jeffco 9700 fuel tank sealer works well because it has better chemical resistance than paint. If you want additional fire protection, use intumescent paint which will protect the structure and prevent burn-throughs for a certain amount of time if subjected to fire.

Spinner Installation

You have a blank spinner and need to make the cutouts for the propeller blades. Assuming you have a two-bladed propeller, the first step is to find two points exactly opposite of each other on the spinner dome. Take a piece of paper and trace around the spinner dome. Cut the circle out and fold exactly in half. Make a mark on each end of the fold line. If you have a three-bladed propeller, fold the circle a second time about 90 to the first bend. Where the two bend lines intersect is the center of the circle. From this center, draw three lines spaces 120 apart. This in where the prop blades will be. Tape the circle to the floor to prevent it from moving. Place the spinner dome back on top of the circle and transfer the marks to the spinner dome. Ask the manufacturer of your prop for a full-size cross section drawing of your prop at the radius of the spinner. If you cannot get it from them, go to the Sensenich website. They have placed some full sized cross section drawings there for several of their props. Select one that is close to your prop and print it out. The drawing should have a centerline indicated on it. Align the center line with the mark on the spinner dome and copy the pattern onto the spinner dome. Cut out this shape about 1/4 inch inside the line. Next comes repeated filing or sanding and fitting until the prop spinner closely fits the prop and the aft edge of the spinner perfectly aligns with the aft edge of the spinner bulkhead. Then you can drill the spinner with undersized holes and cleco it to the spinner bulkhead. To test for spinner wobble, place a step ladder in front of the spinner and clamp a sharpie marker to the ladder with the point of the marker was just touching the tip of the spinner. With the spark plugs removed, turn the engine over. If you get a small black dot right on the tip of the spinner, no wobble was present. Then the spinner can be removed and the holes are opened up to their proper size.

Composite Baffling

Baffling for an engine does not necessarily have to be made from aluminum. There are good reasons for making it from composite, if you don't mind working with composites. Aluminum has the disadvantage that it corrodes and cracks from vibration. Composite does not corrode or crack and should be more durable and potentially lighter if it is the same thickness as aluminum. It can also be shaped right on the engine and if done right, the gaps and air leaks can be kept very small with a minimum amount of sealant.

Divide the locations for baffling on the engine into areas you think you can build it in one piece and still install it. Generally the fewer pieces the better because there are fewer opportunities for air leaks and less fasteners needed. Sometimes you may find that it is too difficult to install a large piece and you have to divide it up in multiple parts.

Protect your engine with plastic sheets and tape and use cardboard to mock up the supports for the baffling layup. With composite you are not limited to two-dimensional bends, but can use compound curves and continuous pieces with fewer gaps.

Selecting a resin with a high Tg (glass transition temperature) is preferable but not essential. Most laminating resins are suitable for the job because it is not that important if the resin loses a bit of strength from the heat in the engine compartment and softens. If you use glass as fibers, you will likely see the resin turn somewhat yellowish or brownish from the heat. Yellow is ok, but if it gets too brown, it is an indication that it got too hot. You can paint the baffling with high temperature (reflective) paint to enhance the appearance. Alternatively you can use carbon fiber, which is already black and will not look any different when it gets hot. Carbon fiber should not be in direct contact with aluminum because it will cause the aluminum to corrode. When you install carbon baffling, apply a thin film of high temperature silicone between the carbon and the metal.

Installation of Baffle Seals

The best material to use as flexible baffle seal for closing a gap between the baffling and the cowling is reinforced silicone sheet (available from McMaster Carr, for example). It is suitable for high temperatures and does not age and stiffen like rubber. In most cases, 0.625 inch (1/16 inch) thickness is sufficient, but the material may be too flexible to bridge larger (>0.5 inch) gaps. The thicker it is, the stiffer it is and less likely to fold over from the air pressure. Unfortunately the silicone sheet is quite heavy and going to the next higher thickness can easily add half a pound of unnecessary weight. To reduce folds or the need for cuts and therefore air leaks, install it with some slight curvature towards the high pressure side of the engine compartment. To do this, start by drilling the rivet holes in the baffling but not in the silicone sheet. Start at one end by attaching the seal with the first rivet. Before drilling the hole for the next rivet in the seal, pull on it to stretch it slightly. This will cause the seal to curve a bit. Drill the next hole and set the rivet it while it is still stretched. Continue in this manner progressively until the whole seal is attached. It should not have any wrinkles and be slightly curved towards the pressure side. Aluminum pop rivets work well for attachment. If possible, the rivet orientation should be as shown in the sketch. If you have to turn the rivet the other way round, use a washer on the seal side to increase the area and reduce the chance of the rivet pulling through the soft seal.

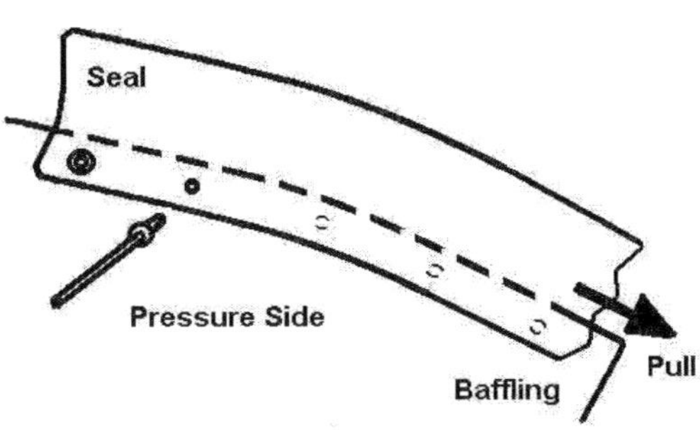

Oil Access Doors

An oil access door is usually located in the high pressure upper plenum of the cowling. This means that in flight there is a lot of pressure on the door, pushing it out. Any play or flexibility in the door itself or hinge and latches will cause it to bulge out or lift up a bit. Some of the air intended for cooling the engine will leak out through these gaps and cause extra drag. The best way to avoid this, if you have the room under the cowling, is to make the oil access door swing inward. That way the pressure pushes against the door in flight and causes a very positive seal. You also eliminate the visible knuckle of the piano hinge.

To build the inward swinging door, the flange will need to be installed on the door rather than on the cowling opening. If necessary, reinforce the skin of the cowling at the opening to stiffen it. If you are modifying your old door, simply remove the existing hinge, tape the door in place from the inside of the cowl and lay up a flange over the existing door on three sides. The side where the hinge will go does not get a flange. Do not forget to put release tape over the cowl where you are going to lay up the fiberglass. Lay up the glass over a larger area than the existing opening to allow for the flange width. You can make a new door in a similar manner. Do the layup over a larger area as it is described in the chapter on access panels.

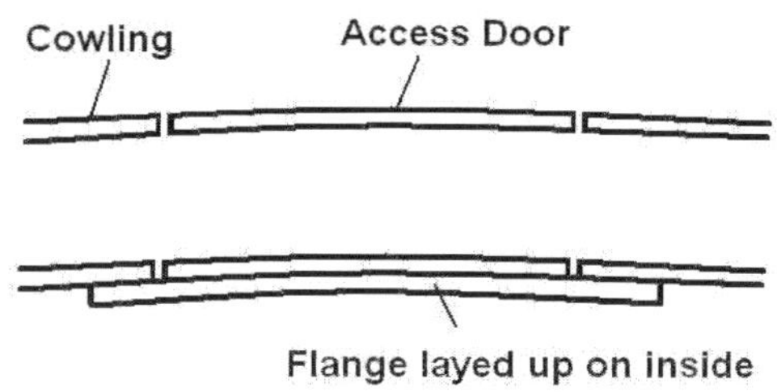

Mount the hinge to the inside of the cowl with rivets. In this installation the door is held closed with a spring. No latch is needed but you could also use a magnet as the latch. If you have no means to hold the door closed, you may find that the door will be in the down position while taxiing but closes automatically once the airplane starts moving fast enough to build up pressure inside the cowling. Some people might like this arrangement. You save the trouble of opening the oil access door after landing to let the hot air escape after you park. If ground cooling (with the engine idling) is marginal for your installation, it is better to have the door closed on the ground.

To install the spring, which was taken out of a hand clamp, cut one knuckle out of the piano hinge and then attach the spring by sliding the hinge pin through the already formed center hole. If necessary, bend the ends of the spring up so that there is enough pressure to close the door. If the door does not close tight on the ground, it will by the time you get airborne.

The Hidden Hinge

Usually hinges on doors which open to the outside will also show on the outside. If you would rather avoid this and prefer the hinge to be invisible from the outside, use a hidden hinge. You can buy the hidden hinge as an assembly or make it yourself. The heart of the hidden hinge is a uniquely shaped arm that allows the hinge to be mounted completely out of view on the underside of the cowl. A spring may be included that loads the hinge in such a manner that forces the oil door "open" when it is unlatched.

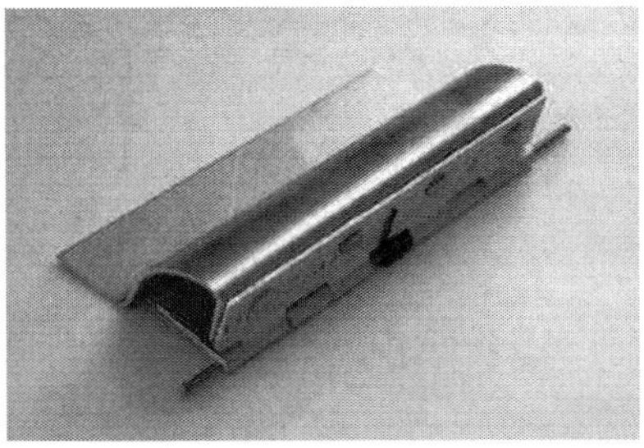

The hinge consists of four components: the specially curved arm, the hinge base, a spring, and the hinge pin. The hinge base and arm come already riveted together and the hinge base is the extruded variety. The hinge should be first mounted to the cowl, then fit the access door to it. If necessary, use a shim to end up with the door flush on the outside. The right photo shows the hinge holding the oil door open, and you can see how the curved arm allows the hinge to be mounted to the inside of the cowl.

Securing Cable Ends

Simple push-pull cables without a fitting at the end of the guide can be attached more securely with a modified clamp. With the standard clamp, the cable can still slide through (left picture). The modified clamp has a groove which grips the cable tightly. This can easily be done with a plain screw driver. Fabricate a die with a groove from a rod of the diameter of the cable. Insert this die in the clamp with the groove centered in the clamp. This will prevent the clamp from simply crushing when you hit the screw driver with a hammer. The resulting groove in the clamp will hold the cable firmly in place once the bolt is tightened.

Oil Cooler Door

Unless you only fly in hot weather, you will notice in winter that the oil temperature will barely creep to the required minimum temperature before take off and in flight it will stay well below the temperature needed to boil off water even though the rest of the engine is reasonably warm. To raise the oil temperature, you can close off part or all of the airflow through the oil radiator. You can build an oil cooler door which is opened or closed by a cable. Even though some engines are equipped with a thermostat (Vernatherm valve) which regulates the oil flow through the radiator according to temperature, this may not be enough. To prevent the oil from congealing in the cooler, there is always some oil flowing through it. Closing off the airflow can be much more effective. A hinged door on one side with a push-pull cable to the cockpit allows you better control over the temperature. Depending on the size of the door, the speed of your airplane and the design of the door, there may be a lot of air pressure on the door. Regular push-button cables may not be able to hold that force. It is typical for such a cable to hold about 15 lb in the knob-push-in direction, while it may be twice that much in the knob-pull-out direction without unlocking it. If you expect to need more holding force, use a locking cable with a handle which rotates to lock. Do not close off 100% of the airflow through the radiator. In most cases, covering 80% to 90% of the radiator area is sufficient and raises the oil temperature 20 F - 40 F. To reduce the load on the door from air pressure, the baffling should be raised to shield the door from the full pressure in the open position (right picture).

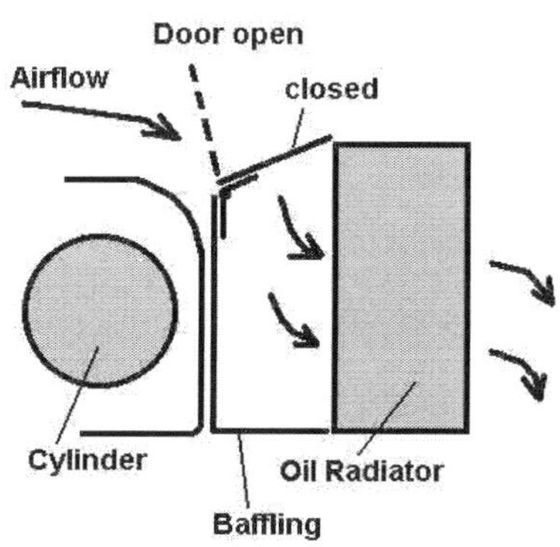

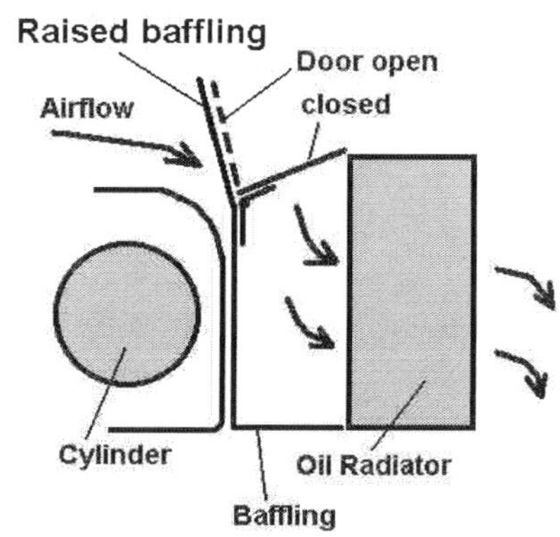

Gaskets

Gaskets in the induction system can be the cause of considerable power loss in flight.

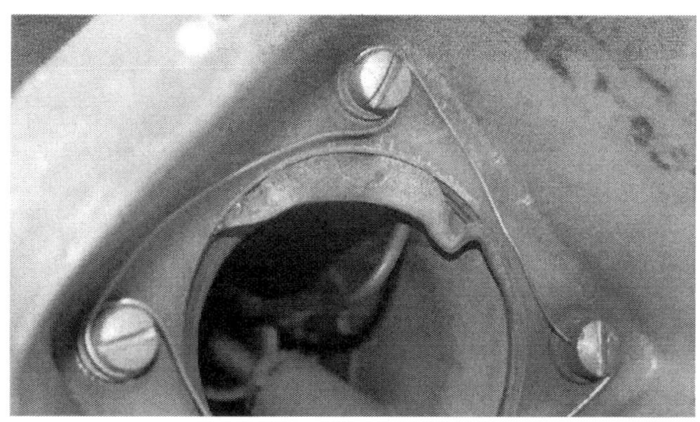

The pictures show why. A strong vacuum can deform the gasket and suck it in to partially block the airflow. The effect is the same as if you partially closed the throttle, with the difference that you cannot open it again in flight. If you have an unexpected power loss, possibly in combination with an overrich mixture, check all gaskets and replace them if necessary. This can happen especially if the pressure on the gasket is uneven because of poor fit of the mating surfaces.

Firewall Insulation

Even though a steel sheet will protect the airframe from a fire in the engine compartment, it will not provide any significant thermal insulation. Even during normal operation the engine compartment and the firewall can become uncomfortably warm. Many builders like to add an insulation blanket to their firewall assembly to reduce the heat transfer and to protect the structure behind the firewall from the direct effect of a fire. A common firewall assembly consists of a steel sheet, an insulation blanket and the airframe structural bulkhead. To keep the insulation blanket from being crushed and therefore rendered ineffective at firewall penetrations where fittings or bolts are tightened, it is necessary to use solid spacers. A spacer for this purpose should have the same thickness as the insulation blanket. It should be made from a flame-resistant material and it should not compress when the bolt is tightened. It should not be made from metal, because it would transfer the heat too rapidly. The most suitable material is a phenolic. It can easily be drilled and cut into the desired shapes. The spacer should be slightly larger than the washer or bolt head used in that location. If it is a structural bolt like an engine mount, it should be at least twice the diameter of the largest washer used in that location. When you assemble the firewall, prepare all the spacers you need and glue them to the airframe bulkhead with instant glue to keep them in place. Cut out the insulation blanket around the spacers and also glue it in place. Then you can mount the steel sheet on top of this, provided you have already cut the holes you need through all the components.

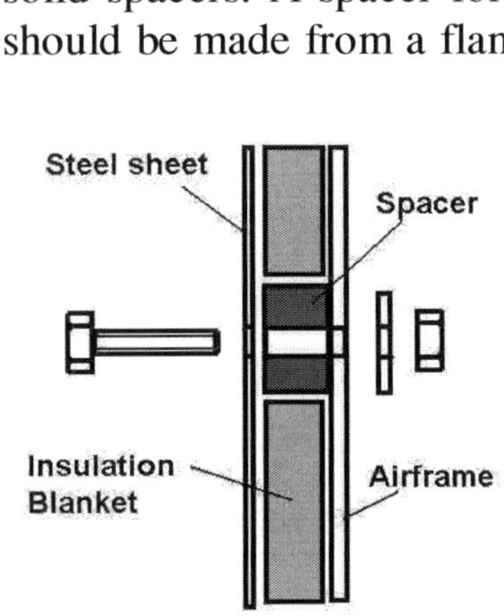

Ensuring Your Fuel System is Alcohol-Proof

More and more pilots who use auto gas discover that their favorite gas station has added alcohol (ethanol) to the fuel. In most cases this will mean you cannot use this fuel in your airplane. If you fly an experimental, you can still use it, but you should be aware of a few things. Most importantly your engine needs to be able to digest this new component. If you ask the engine manufacturer, the most likely answer you will get is: "Don't use it." They do not want the liability. If your engine is an automotive conversion, there should be no reason why you cannot use automotive fuel. You need to make sure all components of your fuel system are compatible with automotive fuel. The easiest way to do so is to use only automotive components. Unless you have a welded metal tank, the tank sealant may be a problem. Most fuel tank sealants have only limited resistance to alcohol and will deteriorate sooner or later. If you have metal integral tanks, you have to use an alcohol resistant sealant. Chemseal PR-1750 is the only one I know of that meets that requirement. It is expensive because it has a short shelf life (unopened 9 months) and is made to order with a corresponding lead time of about 4 weeks. If several builders get together and order a larger batch, it may bring the price down a bit. If you have composite integral tanks, follow the instructions in the plans for sealing it. Most likely they will call for sealing the surface with Jeffco sealant, a gray stuff which cures to a hard surface with excellent chemical resistance. Epoxy and vinylester resins are also quite resistant to alcohol, but will absorb some over time. Unless your tank walls have any kind of porosity, you should be ok. Problems tend to show up after several years of use of fuel with alcohol, not right away. I have operated an airplane with a welded aluminum tank, automotive fuel hoses and a Volkswagen engine for many years (>12) without any real problems on automotive fuel with 10% alcohol.

A few minor issues I came across have to do with water. One thing to keep in mind when using auto gas is its affinity for absorbing water. A lot of water in the fuel reduces the octane rating, even though you cannot see the water. As long as the fuel looks clear, it should be good to use. If it looks milky, do not use it. If you let this fuel sit for a while, the water will collect at the bottom and can be drained. If you have any doubts, reduce the power setting to prevent engine damage. The water in the fuel can create more ice in a carburetor even on a dry, fairly warm day. If your engine starts running rough, try the carburetor heat first. As you climb to higher altitude, the temperature drops and the water may separate from the fuel. Hopefully the sump in the tank and the gascolator are large enough to catch this water before it can reach the engine.

Two-Stroke Engine Operation

As you probably know, two-strokes have a reputation for failing more often than four-stroke engines. Often this is because they were operated in an unsuitable way, and many problems can be prevented if the installer and operator are aware of a few facts about these engines. Two-stroke engines require installation and operation (tuning) in a specific way which is slightly different from the average four-stroke engine. The differences are small but the consequence of not following the rules is far less forgiving. Usually the engine manufacturer makes no attempt to educate the user on how to run their engines properly. In their manuals they tell you the operating limitations of maximum rpm, EGT and the CHT range and that's it. You will hardly get help on how to select the correct propeller for the speed range of the aircraft and how strongly the load on the engine affects its tuning (Rich/Lean). Running the engine with the wrong mixture for the operating condition happens to be one of the most common reasons for the engine seizures. Two main reasons cause engine seizure:

- High temperature difference between the cylinder and the piston which will result in insufficient clearance between the two and causing too much friction, melting the side of the piston and smudging it onto the cylinder wall.
- Lack of lubrication, creating basically the same result.

What prevents the seizure under normal running condition is proper lubrication and temperature control. The engine is lubricated by the oil mixed with gasoline, although injecting the oil is the better way, when done properly. The piston is cooled by the fresh, cool air-fuel mixture entering the crankcase. Its temperature is also influenced by the combustion temperature which is quantified by EGT (Exhaust Gas Temperature). It needs to be kept low enough by keeping the mixture more rich, that is on the cooler burning side with excess fuel. Leaning has two adverse effects. The EGT goes up and because there is less fuel drawn into the cylinder, it also gets less oil for lubrication. So anytime the mixture is leaned the risk of engine seizure is increased.

Two-stroke engines do not have intake or exhaust valves, which means the intake of fresh air-fuel mixture is controlled by tuning and the load on the engine. In a four stroke engine, the fuel-air mixture does not change substantially if the load on the engine is decreased, as it typically happens during take-off. The condition of the airplane with a fixed pitch propeller standing still at the end of the runway with the engine opened up to full throttle produces the highest load on the engine. Once the airplane picks up speed, the load is reduced and you will see rpm increase. Tuning is affected by the induction system, exhaust system, piston porting, carburetor setting, and the load on the engine. All these factors must be in harmony, and you cannot, for example, make up for too low engine load (which creates a lean condition and high EGT) by increasing the jet sizes to lower the EGT. Change any one of the systems to make up for a deficiency in another system is unlikely to improve the overall condition. You may manage to keep the engine from seizing that way, but fuel consumption will be high, temperatures will be odd, peak power will be less and carbon will build up faster.

Here is a typical scenario which could cause a two-stroke engine to seize. The pilot is descending in a shallow descent at a relatively high airspeed and throttle partially open (about half way). The mixture will lean out significantly, but it does not necessarily show up as too high EGT because the engine is not really producing much power. Remember it is not necessarily over the limit EGT but rather the difference between the piston and cylinder temperature which causes problems. All three bad conditions are met here, combustion (piston) temperature is high, cylinder temperature is low (not much power produced and a lot of airflow) and low amounts of oil entering the engine. Even an engine that is otherwise well tuned can seize this way. It is rather recommend descending at only slightly above or at idle rpm at slow airspeed to reduce the strain on the engine.

Another typical problem is when the static rpm is too high which is usually caused by pitch on the prop being selected too low. Many people seem to think that if the book for their two-stroke engine states maximum power is obtained at 6500 rpm, that this is what they should set it for on the ground (static rpm). It is correct that you obtain maximum power at the maximum rpm, but these numbers are way too high and will very likely cause very lean mixture conditions in flight. When the aircraft is in motion, the propeller unloads and the rpm would increase past the limit if the throttle is not pulled back. This will be accompanied by high EGT. If the pilot does not pay enough attention to these warning signs, he may not realize how dangerous this really is until it is too late. The static rpm should be set well below the maximum rpm and the optimum depends largely on the speed range of the aircraft. Generally the higher the cruise speed, the lower must the static rpm be set even if that means lower rpm and power in climb to be able to keep sufficient load on the engine in cruise. Only on very high drag aircraft with a narrow speed range can you set the static rpm higher so it gets close to maximum rpm in the climb as you will not unload the engine on such an aircraft even when you level off for cruise.

Another potentially bad combination is using very large (in diameter or blade area) propeller for a given engine, in which case the propeller pitch must lowered considerably to bring the engine up to the acceptable rpm. In this case, even when the static rpm is rather low, the engine load will start dropping off very rapidly as soon as the aircraft starts gaining speed. Only very slow aircraft can use such a combination successfully, which basically are operated with the throttle wide open most of the time like a hang glider tow plane or a powered parachute.

If you operate a two-stroke engine with the correct propeller, exhaust and carburetor jetting, the engine should perform reasonable well for a long time.

Manual Mixture Control for Two-Stroke Engines

Pilots who fly with two-stroke engines are most likely familiar with the frustration of maintaining the EGT's in the proper range with temperature and altitude changes. Most carburetors have a fixed mixture setting. To change the mixture, you have to re-jet it. To address this problem, you can install a manual mixture control. The concept is based on controlling the air pressure to the float bowl. Normally the air pressure in the float bowl is ambient pressure. The higher the pressure is in the float bowl, the more fuel will flow through the jet to the venturi, thus enrichening the mixture. The opposite is also true. The lower the pressure is, the leaner will be the mixture. Manual mixture control actually controls the air pressure in the carburetor float bowl. One example is the HACman mixture control. It is a simple installation, consisting of a 20-turn valve (vernier control) that requires a half inch hole in the instrument panel and the rest is the controller, fittings, plastic tubing from existing float bowl ports, venturi port, and adding a ambient pickup hole in the air filters.

The HAC unit, which is part of the basic HAC kit shown in the picture, has two chambers separated by a diaphragm. One chamber is completely sealed (except during calibration) and air density within remains constant. The other has airfilter, float bowl, and venturi connections, so air density on this side is variable. It is in this chamber, that ambient pressure can be reduced by feeding it to the venturi via a connection on late model carburetors. The amount of ambient reduction is controlled by a tapered needle which changes position with deflection of the diaphragm. Think of the diaphragm as a flexible wall between two chambers that allows the sealed side to expand or contract as changes in ambient pressure occur on the other side. The reduced (from ambient) pressure gets routed to the float bowl(s) via their vent lines with a resulting leaning affect. If the supply of low pressure is shut off, float bowl pressure returns to ambient, and the mixture goes as rich as the jets will allow. With HAC installations, standard jetting starts out several steps richer. This is why the default, or failure mode is typically toward the rich side.

The system allows the pilot to adjust the mixture to compensate for altitude and density changes or adjust for different loads to maintain the best EGT's. The same concept can also be used with 4-stroke engines which are not equipped from the factory with a mixture control.

Synchronizing Carburetors

One of the cheapest and most accurate tools you can have for balancing dual carburetors is also easy to build. It is a differential manometer (vacuum gauge). You should be able to get the materials to build for less than $10 (or you might already have them).

Here is the shopping list for the Carburetor Synchronizer Tool:

- 20 feet of clear plastic (vinyl) tubing - inside diameter big enough to slip on the vacuum nipple of your carburetor.
- A yard stick or 3 foot piece of wood.
- Clear packaging tape.
- Two short nylon zip-ties
- A small amount of used oil, transmission fluid or colored water, just about anything will work

To build the balancer, fold the 20ft of vinyl tubing in half and mark the center point. Lay your yardstick down flat on a convenient work surface. Place the center point of the tubing at the bottom end of your yardstick. Carefully run the tubing up each side of the yardstick, making sure that the tubing has a smooth, non-kinked bend at the bottom. Use the clear packing tape to fasten the tubing in place on either side (left and right) of the yardstick. Thread the zip-ties through the hole at the top of the yardstick and snug the left and right side tubing to the respective sides of the stick with the zip-ties.

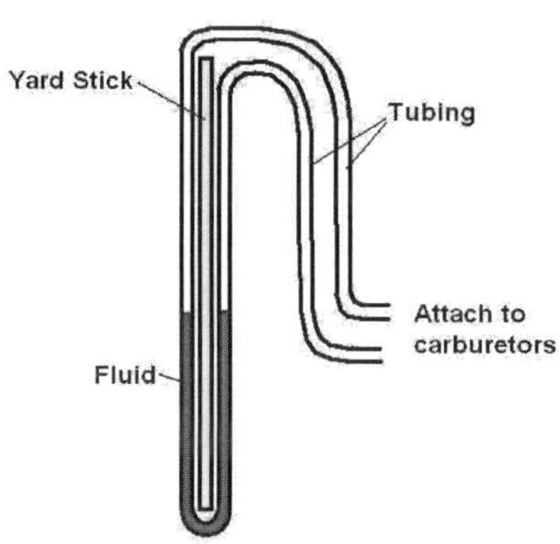

Fold a piece of tape around each end of the tubing like a little flag and mark the left side with an "L" and the right side with an "R" using a magic marker. Put one side of the tubing in the container with the fluid of your choice and, using the other side of the tubing like a drinking straw, suck fluid about three feet up into the tubing. Maintaining suction for a second, pull the tubing out of the container and then raise the ends of the tubing above the top of the yard stick. Temporarily fasten both ends of the tubing high enough that the fluid drains down to the loop at the bottom of the yardstick. You may need to let it sit for a while so that all the bubbles work their way out. If everything has gone according to plan, you should now have a nicely graduated rule with a thermometer-like tube running up both sides, with the "thermometer" fluid about half-way up each side at identical heights. Cut a piece of black electrical tape just long enough to cover the width of your ruler and use it to mark the height of the fluid. Your super-accurate Carburetor Synchronizing Tool is now ready to use.

Preparation for Synchronizing the Carburetors

Warm up the engine to its normal operating temperature and shut it down. Loosen the throttle cable adjuster lock nuts and slack off the throttle cable adjusters until there is absolutely no tension on those cables - they should be loose. Disconnect any springs which would pull the throttle away from the idle position. Remove the balancing tube between the two intakes if your engine has one and plug the holes. Disconnect the vacuum tubes that run from the carburetor nipples to the air filter (or remove the blanking screws or plugs from the vacuum ports, or whatever your carburetors have) and connect the right and left Carburetor Synchronizing Tool tubes to the respective nipples. Secure the airplane with chocks and tie-downs. Pick up a tool that fits the carburetors' idle-stop screws and start the engine. Be very careful to stay away from the prop.

Synchronizing the Throttle Stop Screws at Idle

With the engine at idle rpm, unless you are incredibly lucky, the fluid will begin to move in the tubes, rising on one side of the ruler and dropping on the other. The fluid rises on that side of the engine where the throttle butterfly valve is more closed, which raises the intake vacuum. Gently turn that side's carburetor's throttle stop screw in (clock-wise) until the fluid movement stops, then just a touch more, until the fluid moves back toward equilibrium. As the fluid approaches the balance point (which you have marked with the black electrical tape), turn the throttle-stop screw counter-clockwise and stop the fluid movement right at the balance point. Now check the tachometer for idle speed. If the idle speed is within 100 to 200 rpm of what your engine manual calls for at idle, you have the throttle stop screws perfectly balanced for idle speed. If your idle speed is too high or too low, carefully adjust both throttle-stop screws until the idle speed is within the proper range and the fluid remains stopped at the balance point. Check to be sure you still have slack in both throttle cables. If so, the idle throttle stop screw synchronization is complete.

Synchronizing the Throttle Cables

With the engine stopped, you can now synchronize the throttle cables. Both throttle cables should still be slacked-off and their lock-nuts loose. Carefully turn the throttle cable adjusters counterclockwise until most (but not all) of the slack is removed from the cables, then spin the throttle-cable lock nuts down (clockwise) until they just barely seat. Re-check - there must still be some slack in both cables. You should still be able to move the cable adjusters and lock nuts easily with your fingers. Start the engine. Idle speed should be the same as you have originally set it.

With the engine running, slowly push the throttle in until the engine speed picks up to 3000 RPM or about half of the maximum rpm. Hold it there and check the fluid levels. If the fluid is moving (as is likely), turn down (clockwise) the throttle cable adjuster on the low side of the fluid until it stops, then moves back toward equilibrium, while holding the RPM at 3000. As the fluid reaches the balance point, turn the throttle cable adjuster counter-clockwise and stop the fluid movement as it reaches the black-tape-marked balance point. Snug the throttle-cable adjuster lock-nuts and re-check the balance by pushing in the throttle. If the fluid does not move from the balance point as the RPMs come up, shut off the engine and tighten the lock nuts. If the fluid remains balanced, you now have perfectly synchronized carburetors. Disconnect the Carburetor Synchronizer Tool, reconnect the engine's vacuum hoses (or replace the vacuum port screws/plugs), reinstall the balancing tube and go for a test flight.

Canopies

Blowing Bubble Canopies

There are several methods to form canopies but only for one of them you do not have to fabricate an accurate mold. Bubble canopies are blown from a flat sheet of plastic and a template which is made from plywood (3/8 inch or thicker). The template defines the basic geometry of the canopy. Two different templates are shown in the left picture.

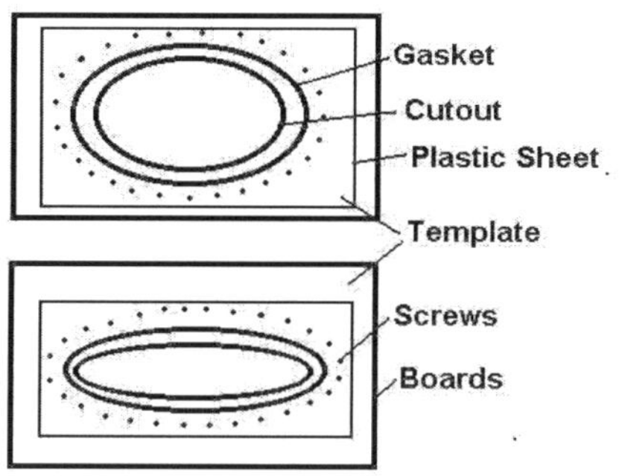

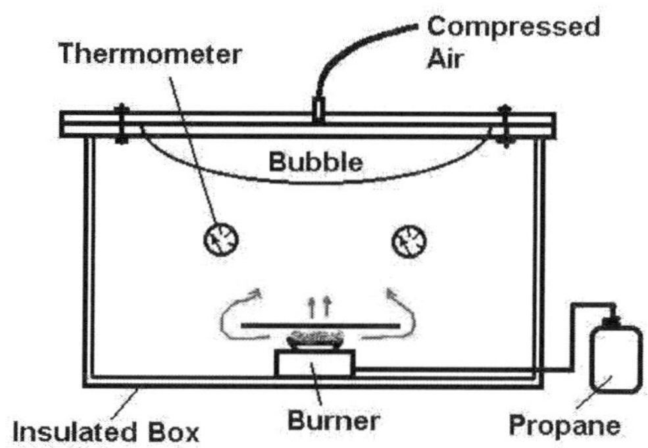

In order to form the material, it needs to be heated. Lexan for example requires a forming temperature of 350ºF. Even though there are several different materials which could be used, Lexan is the most suitable from the quality point of view. It has the advantage of being strong and stable. The thickness of the plastic should be 0.06 inch to 0.09 inch. The thicker it is, the more stable it is, but also heavier. The thinner material would be more suitable for a small canopy (for a single seater for example). The flat sheet of plastic is sandwiched between the template, and a thicker (1/2 inch or thicker) sheet of plywood on top. The sandwich is fastened together with drywall screws. A felt (or other high temperate material) gasket is placed under the screws to provide a good seal between the plastic and the backing plywood. The seal is very important.

You will need to build an oven appropriate to the size of the canopy you want to blow. Aluminum sheets and angles can be used to build the structure. Then it needs to be insulated with high temperature glass insulation. The heating elements will need to be inside the oven. You also need one or more thermometers to check the temperature. Insufficient air circulation inside the oven can create hot spots. To see inside the oven, provide a window and a light source. You can use a large propane camping stove as the heat source (propane tank outside the oven), with a metal hose to feed the propane into the oven. A large sheet of metal over the burners disburses the heat to warm the plastic sheet more evenly.

Heat the oven slowly to the prescribed temperature, although some experimentation may be needed. When the plastic sheet starts to buckle the temperature is almost right. When it starts to sag slightly, you are ready to blow. A standard 10 gallon compressor tank with a hand operated valve is all that is needed to provide pressure. To measure how deep the bubble is, provide markings on the side or a "dipstick". Blow the bubble carefully, watching for any potential blow outs. If you make the bubble deeper than the template is wide, it will balloon sideways, over the sides of the template, and will be hard to remove from the template. If hot spots are caught early, unhook the air hose. The bubble will deflate partially. The hot spot can be fixed, and the bubble can be re-blown (a few times). Be careful not to get the bubble too close to your heat source, as a runaway bubble can happen quickly. Once the bubble has reached a good shape, turn off the heat, and open the oven door, releasing the heat, while using small bursts of air to maintain the bubble's shape. The most common cause for deformed or uneven bubbles is hot spots. If the material is softer in some spots than in others, the result will be unacceptable. Once the bubble has cooled, it can be removed and trimmed. Don't expect your first bubble to be perfect. You may need to practice this a few times before you get the shape you wanted.

Cutting and Drilling Canopies or Windshields

The temperature of the Plexiglas should be at least 70° F / 21° C when you cut or drill it, to reduce chances of cracking. The colder it is, the more brittle it becomes. The Plexiglas may be marked with a fine Sharpie type marker. To cut the windshield, the best tool is an air die grinder fitted with a thin cut-off wheel (abrasive carbide blade). It does an excellent job. Use the slowest speed that will work. **DO NOT** use a jig-saw or saber saw with a saw-tooth blade as this may cause little stress fractures that will later turn into large cracks. You can cut canopy material with an abrasive carbide blade in the jig saw, but make sure you do not have a model that has an advancing blade. That can chip or crack the material. Generally you want only a smooth up and down motion with light pressure applied in the direction of the cut. Again, cut slowly. The grinding action of the cut-off wheel will actually melt the Plexiglas as you cut. You should quickly get a feel for the proper speed to progress. The molten Plexiglas on the edges can be broken or picked off immediately after cutting,

Cut oversize first, then using finer and finer sanding disks, finish the edge to the line you marked. It is a good idea to finish the cut edges with as fine an abrasive as you can get - 400 grit seems to result in the most consistent service life. If the cut edge is left with the cutting pattern marks or coarse sanding marks, those can be crack initiation points when the part is placed in service. As you get to your line and are in the process of doing the final sanding of the edge, make sure to round the corners with about a 0.050 inch radius. Any sharp corner is likely to chip and cause a stress concentration which can initiate a crack even when the overall edge is smooth.

Acrylic sheet can be drilled using commercially available power driven equipment, such as portable drills and drill presses. Drill bits manufactured specifically for plastics work best when drilling acrylic sheet. However, metal working, high-speed twist drill bits can be used with minor bit modification. These metal-working drill bits are made to bite into metal when pushed. If they are used as is, metal-working drill bits will chip acrylic sheet as well as cause other damage. They must be reground in order to scrape instead of biting or gouging the sheet. Follow these steps for modifying standard metal working twist drills to work with plastics:

1. Grind the tip angle (usually a relatively flat 120º) to a sharp 60º - 90º. This sharp angle allows the bit to easily enter and exit acrylic sheet without chipping.
2. Grind the cutting edge, flat, to a 0º – 4º rake angle. This cutting edge should scrape acrylic sheet, not gouge it.
3. Grind away the surface behind the cutting edge to clearance angles of 12º - 15º. This back relief reduces metal-to-plastic contact and heat build-up.

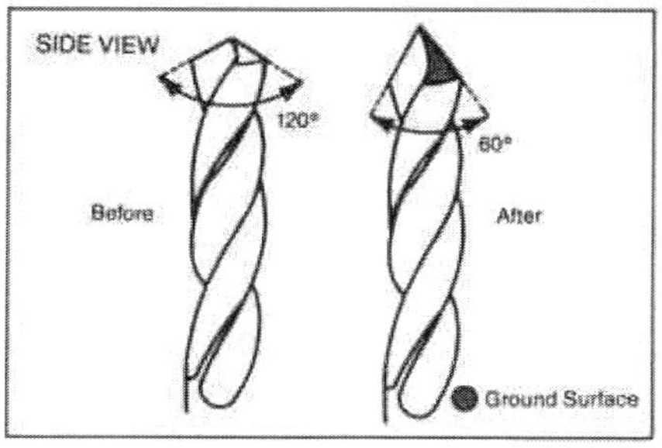

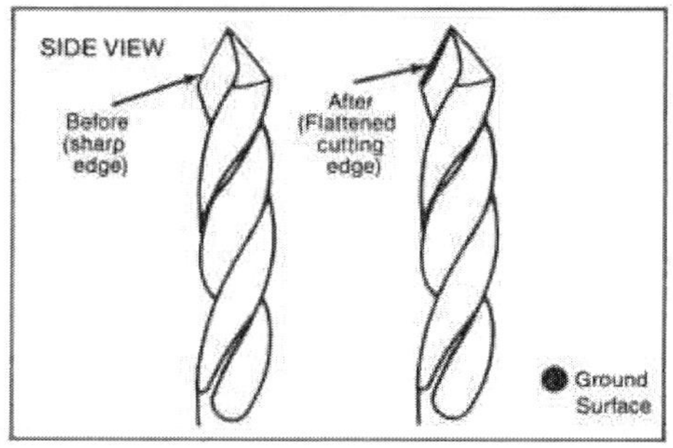

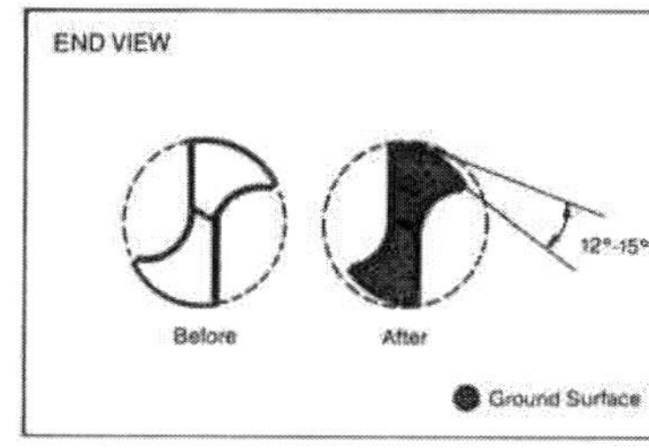

Correctly modified bits operated at proper speeds (see the table below) will create two continuous spiral strips as the bit passes evenly through the acrylic sheet. Allow the bit to enter the material using a slow feed rate. Slowing the feed rate as the bit exits the bottom surface prevents chipping. When drilling a stack of acrylic sheet (1/4 inch thick or greater), use water as a cutting fluid. The water acts as a coolant, helps to reduce heat build up, and creates holes with smoother walls. Make holes for fasteners slightly oversize to allow for expansion and contraction of the acrylic material. Holes that may be subject to force from screws or bolts should be deburred with a countersink. Zero flute countersinks work well for countersinking and deburring holes in acrylic sheet. If a countersink is unavailable, use a drill of larger diameter than the hole, and deburr the rough edge on the exit side of the hole (the side where the drill came through the sheet).

Drill Diameter	**Speed (rpm)**
1/8 inch / 3.2 mm	3500
3/16 inch / 4.7 mm	2500
1/4 inch / 6.4 mm	1800
3/8 inch / 9.5 mm	1200

Do not tighten any hardware beyond making slight contact. Loosen screws a half turn after bottoming.

Clean a dirty canopy with plenty of water and dry with a clean chamois or soft cloth. Light dust can be removed with an air blast or a damp soft cloth. As convenient as paper towels are, do not use them because of their cellulose content which scratches the acrylic. Never use any abrasive cleaners and be very cautious about using any form of solvent on Plexiglas. “Mirror Glaze™” is recommended for cleaning and small scratch removal. A “Micro Mesh kit” will remove deeper scratches. Automotive paste wax may be applied but be sure it is NOT a “cleaner / wax”. Most of those products one can find on the shelf at the local store are “cleaner/waxes” and have varying degrees of rubbing compound in them which of course would create fine scratches or swirls.

Storage of Acrylic and Removal of Protective Coating

Ideally, acrylic sheet should be stored almost vertically in special racks with plywood panels for full support where sheets can lean at about a 10º angle. If stored horizontally, sheets must be clean because the weight of the material can cause dirt or chips to pierce the protective masking, thereby scraping or indenting the sheet. When stacking various sizes horizontally, make sure to put the larger sheet on the bottom to avoid unsupported overhang. The same applies to storing a formed canopy, except that of course you would not put it between two sheets of plywood. Keep acrylic sheet away from heat sources, such as radiators or steam pipes, which can deform or soften the sheet.

Acrylic sheet is manufactured with paper, polyethylene or latex film masking, on both sides, to protect it against scratching during storage and handling. Folding and re-taping the overwrap after it has been opened will minimize warpage which can result from moisture absorption. Avoid placing acrylic sheet near spray painting booths or exposing it to solvent vapors which can penetrate the masking and damage the sheet's surface. Canopies are shipped with the protective film in place. This masking should be left in place during most fabrication operations. It can be removed, if necessary, for intricate detail work or in line bending and thermoforming operations. Usually the masking peels off easily. You can mark the lines for cutting the sheet with the masking in place but marking polyethylene masking requires grease pencils or special felt tip markers.

If you have had the canopy for a long time, you might find that the masking ages and becomes brittle and difficult to remove. Exposure to sunlight or moisture can also make masking removal very difficult. If the sheet will be exposed to the outdoors, the masking should be removed immediately after installation. Old masking with hardened adhesive can be moistened with aliphatic naphtha, hexane, or kerosene for easier removal. One method is to spray it with WD-40, let it soak a few minutes, and then you can peel it away a little bit at a time. If you try spraying it with anything else, first make sure it will not attack your canopy and craze it. Do not use acetone, MEK or gasoline on Plexiglass. Use water or alcohol for cleaning. Wash the oily solvent residue away immediately. Do not use sharp-edged objects, such as razor blades.

Canopy Repair

A crack in a canopy does not mean the canopy has to be replaced. A crack can be repaired. The only suitable adhesive for acrylic canopy repair is two-component acrylic adhesive like Weld-On 40. If done well, the repair may be almost invisible. Start by grinding a grove on one side along the crack about halfway through the material. Use a slow turning tool to minimize heat buildup. Position the surface horizontally otherwise the adhesive will flow from the crack. Tape up the crack on the lower side for the same reason and to keep the pieces aligned. Fill the crack with adhesive. Keep in mind that it will shrink, so build it up a bit above the contour. When it is cured, turn it around and repeat the process on the other side. To remove the built-up material, use a small, hard sanding block and 400 grit sandpaper. Wet-sand the excess material, then follow up with 600 grit, 1000 grit, 2000 grit and acrylic polish. Be careful to remove as little as possible of the original canopy material, otherwise it creates an optical distortion.

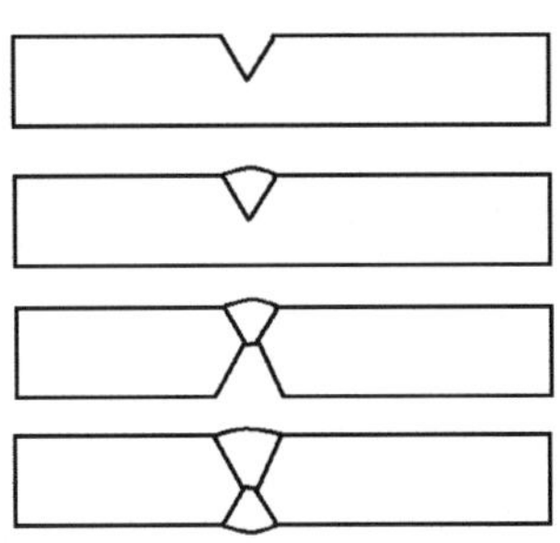

Bonding Canopies

Canopy frames are usually much stiffer than the canopy itself and have a different thermal expansion coefficient that the canopy material. Acrylic plastic shrinks much more than fiberglass, carbon or steel when it gets cold. Depending on its design, either the canopy frame deforms and gaps open up when you climb to altitude or the canopy will be stressed. Any stress concentrations like holes, window openings and fasteners will sooner or later become starting points for cracks. The best way to attach canopies or windows to their frames is by bonding. Not all adhesives bond well to acrylic. My favorite adhesive is commercially available polyurethane window adhesive. It is a flexible sealant that holds up very well over time, sticks to anything, can be painted over

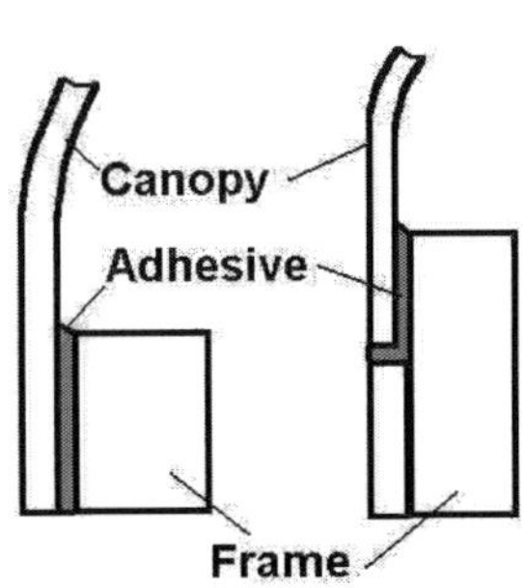

unlike silicone (RTV) and does not cost much. An inch or two wide bonding seam will hold the canopy securely in place. It can be cut with a knife if the window ever needs to be replaced. Because it is flexible, it will decrease the stress on the canopy from temperature changes and eliminates any chance of cracked bond gaps. Another material often used to bond windows is fuel tank sealant. It adheres well to just about anything and is equally strong enough to hold a window or canopy in place. In either case, mask the canopy well against sealant in places where you do not want it because it is very difficult to remove it later with solvents without affecting the canopy material.

Preparing for Flight

Estimate the Weight Paint Adds to an Airplane

You can estimate how much weight that primer and paint will add to your airplane from the area to be painted and the thickness of the paint. Calculate the surface area of your airplane. This is for example the wing area times two (upper and lower surface), plus the fuselage skin area and the tail area times two. Typically the paint thickness is 0.008 inch to 0.015 inch thick (0.2-0.4 mm). Paint weighs about 70 lb/ft^3. If you calculated the surface area to be painted in ft^2, multiply it with the paint thickness, the specific weight of the paint divide it by 12 to obtain the weight of the paint in lb:

Weight of paint [lb] = Surface area [ft^2] * paint thickness [in] * 70 lb/ft^3 / 12

A few examples of what the paint would weigh are shown in this table for typical airplane sizes:

Weight of Paint	Paint Thickness [inch]		
Airplane	0.008	0.012	0.015
4-Seater	22.7 lb	34 lb	42.5 lb
2-Seater	14.8 lb	22.2 lb	27.7 lb
Single Seater	8.9 lb	13.3 lb	16.6 lb

Structural Load Tests

Anytime a new airplane is designed and built, it is important to prove that the structure will withstand flight loads. Even the best analysis cannot take everything into account. For homebuilt aircraft, loads and structural analysis are frequently rather superficial. Defects in the construction process or material substitutions may lead to a weaker structure than anticipated. The only way to find out what you got is to test it. Usually the airplane is designed to withstand higher positive loads than negative loads. Because the lift of the wings acts upwards and the fuselage is pulling downwards with its weight, a static test for the wings requires the airplane to be inverted if weight is used to apply the loads. Horizontal tail loads are usually downwards so it can be tested right side up unless it is a canard, which has to be tested upside down like a wing. During a wing test, it is essential that the upside-down fuselage is supported in a way which spreads out the loads to prevent damage to it. If that is not possible, the wing must be tested without the fuselage. In this example it was possible to test the wings with the fuselage attached.

This Varieze was placed upside down on a steel support structure, so that the two top longerons lay flat and level on top of two strips of PVC foam to spread the load. The wings and canard were installed, the ¼ chord on the wing and canard lower surfaces were marked and a series of butt lines 6 inches apart laid out, full span. These markings were used to position the shot bags precisely at known butt line and fuselage station positions. Then a lot of 25 lb shot bags were piled up on the wings and canard. It was decided to load the canard first, but in order to do that, a bunch of shot bags had to be stacked on each wing root, to prevent the whole plane from nosing over! As you can see from the enclosed photos, there were 6 shot bags stacked vertically starting at the side of the fuselage, and working outboard. They ended up with 8 vertical stacks of six 25 lb shot bags, one stack of two shot-bags, and finally just one shot-bag at each canard tip.

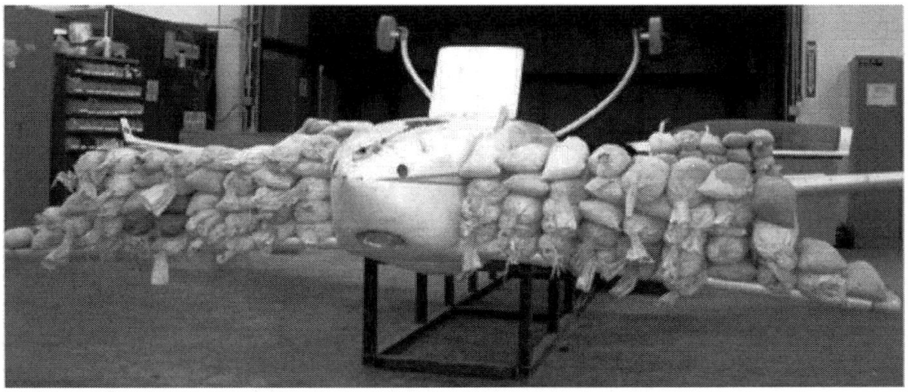

In this test, there was no support structure underneath the loaded structure to keep it from dropping more than a few inches if it failed. In most cases of individual builders it will be sufficient to apply limit load, which is the maximum load the airplane is allowed to see in service. In the case of an airplane manufacturer, a test to ultimate load or failure, whichever comes first, is required. The ultimate load must be at least 1.5 times the limit load.

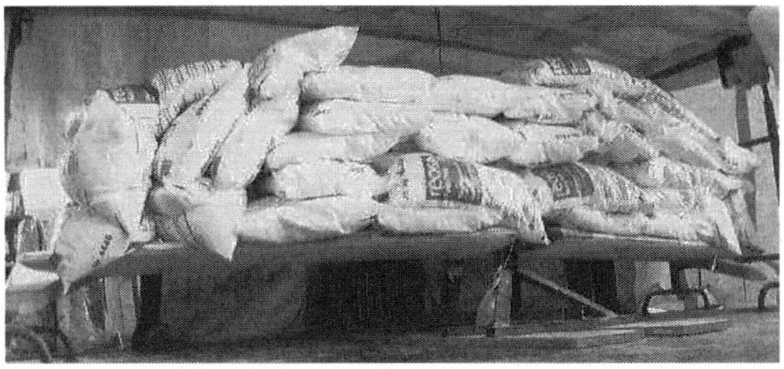

These two examples show wing load tests without the fuselage. In these cases, a fixture is needed to support the wing at the normal fuselage attach points. The more flexible the wing is and the larger the wing span, the higher has to be the support to keep the wing tips from touching the ground when they bend under the load. Whatever you use to apply the load is up to you, as long as it is heavy and a plentiful supply is at hand. Lead shot bags are best because a 25 lb shot bag takes up less space than a 25 lb sandbag. The more weight is needed, the greater can be the bulk until you can hardly reach the top of the pile. In flight, the load is divided between upper and lower surface (pressure and suction). During the load test, all the weight rests on the lower surface. This may require protection of the skin to spread the load and avoid damage from concentrated loads. You can use styrofoam sheets between the wing and the weight for this purpose. If you use a different material, make sure it does not help stiffen the structure and help it carry the load. If the wing has a lot of dihedral or is loaded at an angle, the weight might slide off a slick surface. To prevent this, tape strips of coarse sandpaper, pieces of carpet or rubber mats to it to prevent this.

Before you perform a structural load test, ask the designer what the load should be and how it is supposed to be distributed. Every airplane has different requirements.

A rather unconventional negative wing load test is shown in the following picture!

Final Assembly

When you get close to finishing your airplane, there will certainly be a lot of little things which still need to be done. Make a list of all these things by writing them down when they come into your mind. Check them off the list as you complete them, in order not to forget anything.

Airplane Weight and Balance

To weigh an airplane, you should have three scales, one for each wheel. With the airplane positioned on the scales, it needs to be in level attitude, as specified by the plans or builder's instructions. Read the weight on all three scales and measure the location of the wheels from a datum. It does not really matter where that datum is, as long as it is the same as what is used to define the limits of the CG envelope. It can be the tip of the spinner, the firewall, the wing leading edge or anything else that is convenient. The arms measured from the datum are positive if they are aft of the datum or negative if they are in front of the datum. Use a plumb bob to measure the position of the wheels from their center and mark it on the floor. Once the airplane is moved out of the way, the arms can be measured from the marks on the floor. From this information you can calculate the empty weight and center of gravity.

Include all items in the weighing procedure which you want as part of the empty weight. That should include engine oil, coolant, unusable fuel, brake fluid, seat cushions, wheel pants etc. Can you weigh an airplane if you have only one scale? Yes, but it is more work. You will need to level the airplane first with the scale under one wheel while the other wheels are supported by blocks of wood of the exact thickness of the scale. Take the first reading, the move the scale to the next wheel, but return the airplane to the same attitude as before. Repeat the process with the third wheel. It is important to level the airplane not only longitudinally but also laterally, otherwise you cannot assure that the weight read at the left and right wheel are correct. Once you have the weights and arms, you can enter the numbers in a spreadsheet or calculate it manually. It is really simple. To use a spreadsheet in Excel, type in the following information:

Item	Weight	Arm	Moment
Left Wheel	430 lb	12 inch (Arm 1)	430 x 12 = 5160 in lb
Right Wheel	450 lb	12 inch (Arm 2)	450 x 12 = 5400 in lb
Nosewheel	45 lb	-32 inch	45 x (-32) = -1440
Total	925 lb	= 9120 / 925 = 9.9 in	5160 + 5400 – 1440 = 9120 in lb

The total weight is the sum of the three individual weights. Add up all the moments for the total moment. The CG is the total moments divided by the total weight.

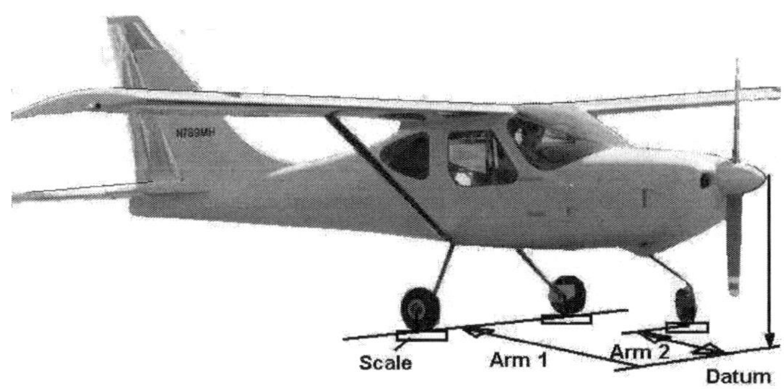

Increasing the Range of a Scale

If you cannot get hold of a scale designed to weigh airplanes or with the proper range, you can adapt a bathroom scale. This may not be as accurate, but can save a lot of money if you otherwise would have to buy or rent scales. Most airplanes are too heavy for a bathroom scale with its typical range of 250-300 lbs. To increase its range and not overload it, build the following extension:

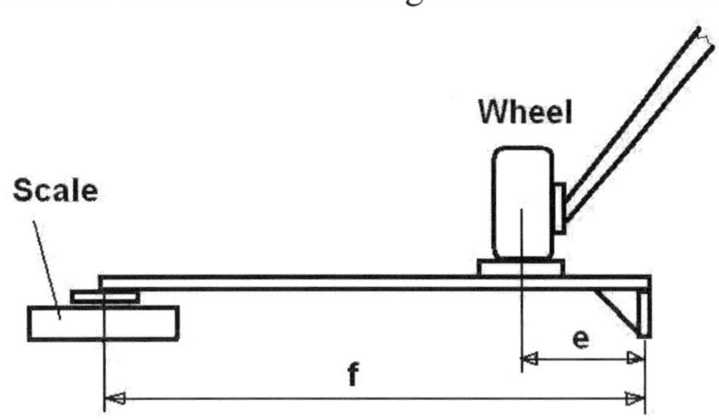

One end of the beam is supported by the scale, the other end is placed on the ground. The beam in between has to be perfectly level, otherwise the reading will be wrong. Place the wheel of your airplane on the beam, but closer to the end supported by the floor. To get the weight of that wheel from the reading of the scale, you need the two distances e and f, as shown in the sketch.

Weight of Wheel = Scale reading * f / e

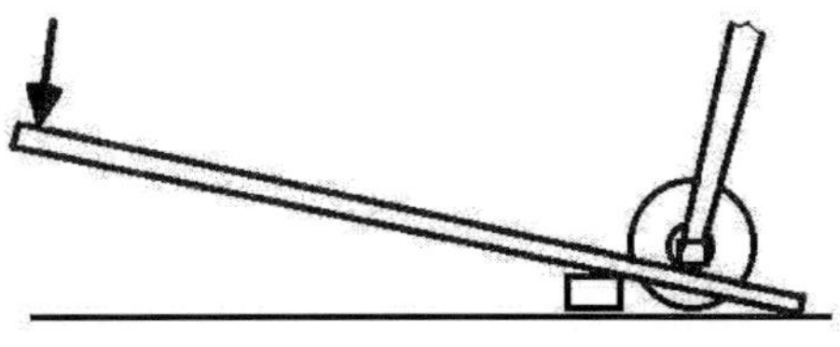

Jacking Points

On many occasions there will be a need to jack the airplane up, for example when you need to change a tire or when you weigh it with scales you cannot roll it up on to. If you planned ahead while you built the airplane, you or the designer have created hardpoints for that purpose. But in many cases this was not considered in the design. One way to jack an airplane up is with a lot of helpers. I have seen that many times. Five people get under the wing and push it up. But what if you do not have five guys handy? Most airplanes have a place at the wheel that is sufficiently strong. This is where a lever can be applied. Take a long sturdy pole and a brick or two and you should be able to lift it by yourself while a helper places a block / scale / support where you need it.

Carbon Monoxide

Carbon monoxide (CO) is a poisonous gas which is part of the engine exhaust gas. The airframe must be sealed to prevent exhaust gas from entering the cabin. Typically the exhaust is discharged underneath the belly. If you have cut back your exhaust pipes too far, this problem is more likely to occur. The cockpit usually has slightly lower air pressure than the air underneath the airplane, especially at high angle of attack. Exhaust gas may be drawn into the cabin during some or all flight conditions unless the bottom is well sealed. Unfortunately there may be seams, gaps or joints along the belly which would allow the gas to enter the cabin. Before the first flight, look closely at the structure which is in the path of the exhaust. Seal all gaps either with high-speed tape or silicone sealant. If you can smell exhaust gas, you most likely already have dangerous quantities of CO in the cabin. Use a CO detector during the initial phase of flight testing until you are sure the cabin is well sealed.

Flight Characteristics and Performance

Reducing Pitch Sensitivity

Some designers seem to be unable to determine the proper center of gravity range for their airplane. I have seen it often that the aft CG limit was selected so far back that the airplane is nearly unstable in pitch or very sensitive in the elevator. Find out in advance if that is a common problem with the model you are planning to build. It may limit your loading options or result in a tendency for pilot induced pitch oscillations. If you have an issue, there are several ways to improve it. The easiest is to set a new limit for the aft CG, moving it further forward. If that is not practical, consider increasing the size of the horizontal tail or increasing the length of the fuselage to move the tail further back. The only problem with this may be that the added or shifted weight of the tail also moves the center of gravity back, which needs to be compensated with more weight up front for balance. Another option may be to change the gearing of the stick to elevator control. If for example you have 2 inches of stick movement for 30 of elevator travel, you have a very sensitive elevator control. It is better to have a much larger stick travel so that small movements of the stick are translated into small movements of the elevator. To do this, you may need to change the length of a bellcranks in the control system. This reduces the overall control surface deflection unless you also increase the total stick travel.

Wedges on Control Surface Trailing Edges

Sometimes it is not all that good to have very thin control surface trailing edges. It is low drag, but it may negatively affect the flight characteristics. The boundary layer, which consists of slowed down air particles close to the surface of an airfoil in the air stream, is very thin at the leading edge and at its thickest at the trailing edge. This can affect the force you feel or rather not feel when you deflect the control surface a small amount. One effect may be a certain hysteresis when you attempt to trim the airplane if the affected control surface is the elevator. In still air the aircraft will trim, sometimes with good stability, but other times it will prove very difficult to trim and will need constant attention. A low stick force gradient may be another result of a thin trailing edge.

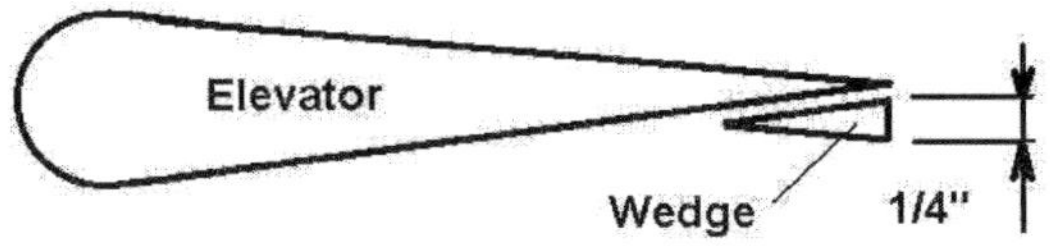

To fix these problems, attach a small wedge to the trailing edge of the control surface. You may need to experiment with the size and location of the wedge or wedges. On an elevator, the wedge will most likely need to be on the bottom surface. The length of the wedge does not necessarily need to be full span, start with 1/4 of the span. If the wedge is too large, it may affect the trim setting. You can adjust the trim, reduce the size of the wedge or balance it by splitting one large wedge into two smaller ones on opposite sides of the control surface (upper and lower surface). Always keep the mass balance requirements for your particular application in mind. Adding weight to the trailing edge may result in a dangerously large hinge moment. A wedge can be attached to all control surfaces, not just the elevator. A temporary, lightweight wedge can be fabricated from balsa wood or foam and can be attached with speed tape. Once you have determined the size, make it permanent from a material suitable to the material you airplane is built from.

Another problem that wedges can help resolve is a combination of large freeplay and friction in the control system that can easily result in PIO's (Pilot Induced Oscillations).

Wing Root Fairing

A fillet between wing and fuselage can reduce the drag and reduce separation during slow flight at the wing root if it is done properly. Aerodynamically best is a fillet which starts with a small radius (1/4 inch) at the leading edge and continuously increases its radius towards the trailing edge. Most builders will probably want to make this from composite material since it involves compound curves. It is easiest if the fairings are bonded to the fuselage. Make templates for the different stations with the shape that you desire. Glue them to the fuselage and fill the spaces in between with blocks of foam.

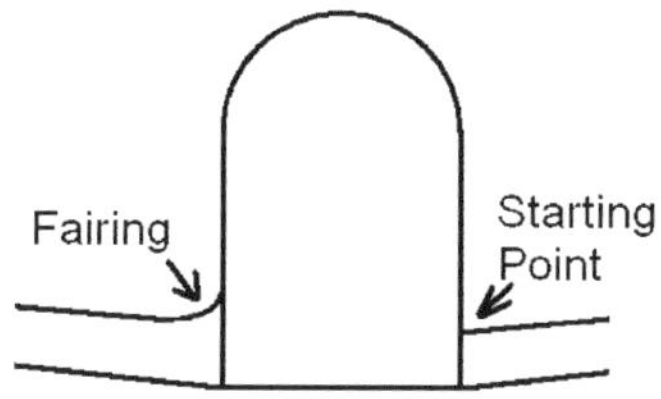

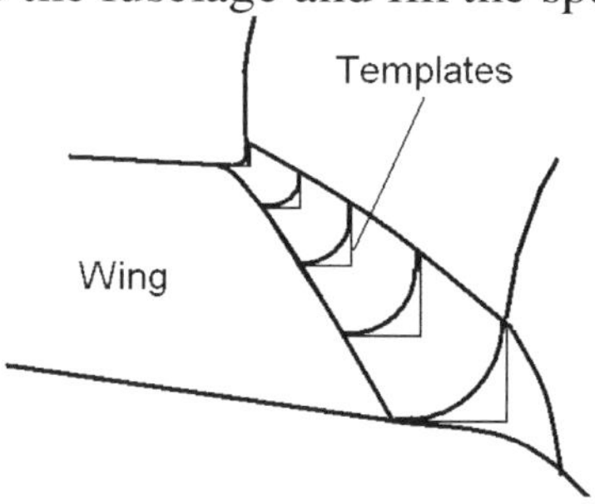

Prepare sanding blocks of different radiuses by gluing sand paper to tubes of different diameters and sand the foam down to the templates. Once you are happy with the shape, lay up one or more plies of cloth over the foam. If the fairing is not supposed to be bonded to the fuselage, release the structure underneath with tape before gluing the templates and foam to it so it can be removed afterwards. Small permanent fairings or fillets of constant radius can be formed with microballons and a spatula to the radius you want.

The builder of this fillet did not use templates but simply sculpted the foam into a shape that looked nice.

Improvement of Airflow in Stalls at the Wing Root

It has been found that low wing airplanes will experience earlier airflow separation at the wing root during slow flight and stalls if the sides of the fuselage intersect the wing at an angle of less than 90º, regardless of the fillet. The picture on the left shows the original rounded fuselage, the picture on the right shows the reworked fuselage shape. A small fillet will also help delay the separation at the wing root.

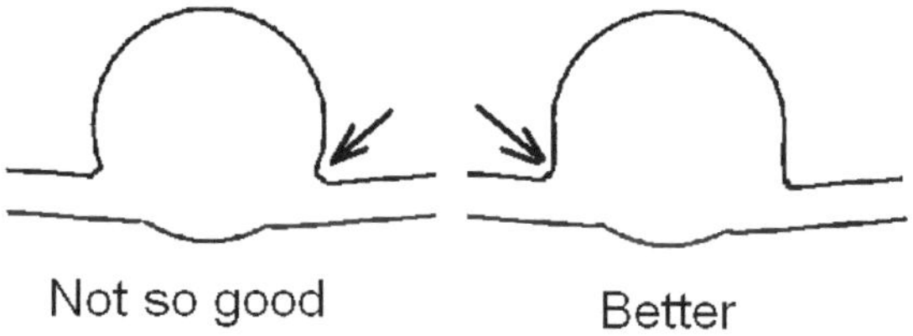

Improving Directional Stability

Most airplanes have a vertical tail which only extends above the fuselage. That causes the slip stream from the propeller, which spirals around the fuselage, to hit the vertical tail at the same angle from one side for the full span of the vertical tail. At slow speed, for example during takeoff, climb or stall, this causes the airplane to yaw, in some cases requiring a lot of opposite rudder input. To reduce this effect, add a small fin underneath the fuselage. The prop slipstream will hit this fin from the opposite side, canceling out part of the effect from the vertical tail on top of the fuselage. It is particularly convenient to do this with a taildragger. I found that a fairing around the tail spring has a lot of beneficial influence on directional stability during takeoff. A lot less right rudder is needed during the initial take off roll to keep the airplane straight on the runway with the fairing, than without it.

Gear Leg Fairings

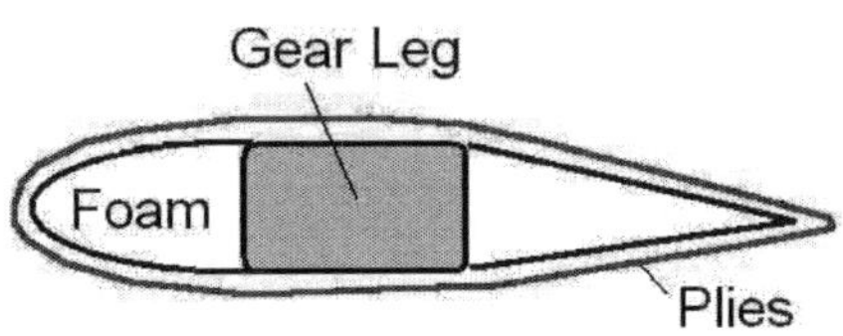

Especially if you have composite gear legs, it is easy to fabricate drag reducing gear leg fairings. Glue blocks of foam or balsa wood to the front and back of the gear leg. The brake line can be enclosed in the foam. To make the brake line removable, bond a conduit in which is slightly larger than the line or hose used. The line can then be inserted or removed through this conduit. Sand the foam to obtain a teardrop shape. Layup two plies of cloth over the shaped foam.

Wing Leading Edges

Besides the selection of the airfoil and the wing planform, nothing has as much influence on the stall characteristics of your airplane as how the contour of the wing leading edge matches the design. Regardless of what the rest of the wing looks like and how docile other airplanes of the same model may stall, here is a good potential for trouble. First of all, the shape of the leading edge should be the same left wing to right wing. If one side is different, it may stall at a slightly lower angle of attack, which means one wing drops, possibly rather suddenly. Even small deviations from the ideal airfoil contour near the leading edge can give unexpected results. The picture below shows in an exaggerated way the typical deviations found on leading edges. This happens mostly with wing skins which are bonded at the leading edge.

Because deviations from the correct contour by as little as 0.040 inch (1.0 mm) can have a detrimental effect on stall characteristics, it is necessary to use templates to look for and correct these deviations. To make a template, you either need a full size plot of the airfoil contour from the designer of the airplane or draw it yourself if you have the airfoil coordinates. Plot the contour on a piece of paper and glue it to a backing surface for rigidity. A leading edge template can be made from cardboard, aluminum or any other sheet-type material. It should extend to about 20% of the wing chord on the upper surface and about 5% on the lower surface The more closely the wing contour matches the templates, the more likely it is to stall straight. Check your contour before and after the wing is painted. Too much paint in the wrong place can change things.

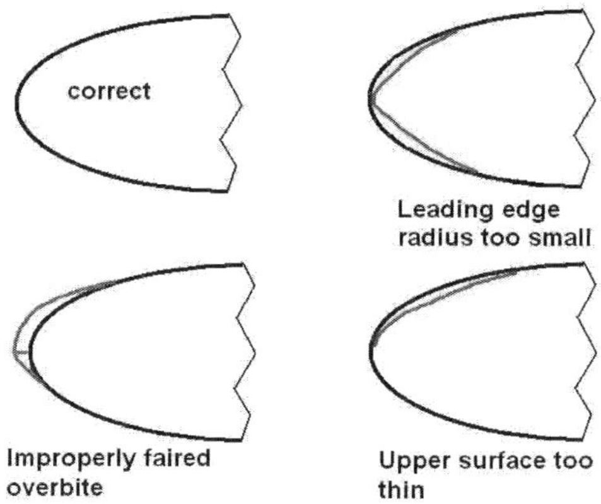
correct
Leading edge radius too small
Improperly faired overbite
Upper surface too thin

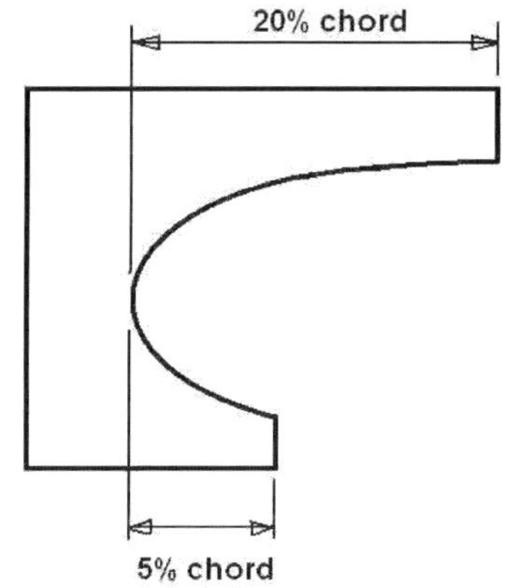
20% chord
5% chord

About the Author

Sonja Englert was born in Germany. At age 16, she started flying gliders, soon adding motorgliders and airplanes to her ratings. While still in high school, she worked as an intern at a German sailplane factory. She studied aeronautical engineering at the University of Braunschweig, where she gained an equivalent of a Masters Degree. During her time as a student she joined the Akaflieg Braunschweig, a group of students who spent their spare time designing, building and flying sailplanes. Sonja participated in building the sailplane SB13 flying wing, which made its first flight in 1988. For the following project, the SB14, she designed the composite wing structure as part of her thesis. Several of her Akaflieg test projects were directed towards investigating drag reduction and flight characteristics of gliders. As a highlight, in 1992, she won the National Cross Country Soaring Championship flying the SB 11, one of the Akaflieg's older designs.

After completing her studies, she started to work as a Design and Certification Engineer for the Daetwyler company in Switzerland, which was developing an all-metal two-place aerobatic trainer. After accomplishing the Swiss and FAA certification, the project was sold to Malaysia. Subsequently Sonja spent some time there to help start the production process and assist with local certification. After returning to Germany she worked as a consultant to develop and design a composite replacement wing for Mooneys, which would allow a Mooney, through extensive use of laminar flow, to fly about 20 knots faster on the same horsepower. Through this project she got in contact with Mod Works in Florida, a small company which modified Mooneys. Although the wing project was discontinued due to lack of funding, she was employed by Mod Works, where she worked on engine installations for several airplane models. By increasing propulsion efficiency and reducing cooling drag, the speed of a Mooney was increased by 15 knots vis a vis the original version, both with the same horsepower. She was further involved with several STC projects and NASA funded research projects.

In 2001 Sonja joined Adam Aircraft to develop the A500, a composite, pressurized twin-engined airplane, with a centerline thrust engine arrangement. She did the structural design of the prototype tail and booms and was responsible for developing the engine installations as the powerplant lead engineer.

From 2002 to 2009 she worked for Lancair/Columbia/Cessna on the development of the C350 and C400 models. She was responsible for the engine installation, cooling and much of the flight testing of the Columbia 400. A large part was dedicated to spin resistance testing and developing modifications to making the Columbia 400 spin recoverable. She currently works as a consultant on various projects in aviation.

Sonja has accumulated more than 4000 flying hours with a commercial airplane and glider pilot's license, with instrument and multiengine rating. She enjoys flying her homebuilt Pulsar and a motorglider.